New Observations

on the

Natural History

of

BEES

by

FRANCIS HUBER

Translated from the Original

To

SIR JOSEPH BANKS, BART.

KNIGHT OF THE MOST HONOURABLE

ORDER OF THE BATH,

A PRIVY COUNCILLOR,

PRESIDENT OF THE ROYAL

SOCIETY OF LONDON,

&C. &C.

THIS TRANSLATION

IS INSCRIBED.

Contents

TRANSLATOR'S PREFACE

The facts contained in this volume are deeply interesting to the Naturalist. They not only elucidate the history of those industrious animals, whose nature is the peculiar subject of investigation, but they present some singular features in physiology which have hitherto been unknown.

The industry of bees has proved a fertile source of admiration in all countries and in every age; and mankind have endeavoured to render it subservient to their gratifications or emolument. Hence innumerable theories, experiments, and observations have ensued, and uncommon patience has been displayed in prosecuting the enquiry. But although many interesting peculiarities have been discovered, they are so much interwoven with errors, that no subject has given birth to more absurdities than investigations into the history of bees: and unfortunately those treatises which are most easily attained, and the most popular, only serve to give such absurdities a wider range, and render it infinitely more difficult to eradicate them. A considerable portion of the following work is devoted to this purpose. The reader will judge of the success which results from the experiments that have been employed.

Perhaps this is not the proper place to bestow an encomium on a treatise from which so much entertainment and instruction will be derived. However, to testify the estimation in which it is held in other nations, the remarks upon it by the French philosopher Sue, may be quoted, 'The observations are so consistent, and the consequences seem so just, that while perusing this work, it appears as if we had assisted the author in each experiment, and pursued it with equal zeal and interest. Let us invite the admirers of nature to read these observations; few are equal to them in excellence, or so faithfully describe the nature, the habits, and inclinations of the insects of which they treat.'

It is a remarkable circumstance that the author laboured under a defect in the organs of vision, which obliged him to employ an assistant in his experiments. Thus these discoveries may be said to acquire double authority. But independent of this the experiments are so judiciously adapted to the purposes in view, and the conclusions so strictly logical, that there is evidently very little room for error. The talents of *Francis Burnens*, this philosophic assistant, had long been devoted to the service of the author, who, after being many successive years in this manner aided in his researches, was at last deprived of him by some unfortunate accident.

Whether the author has prosecuted his investigation farther does not appear, as no other production of his pen is known in this island.

It is vain to attempt a translation of any work without being to a certain degree skilled in the subject of which it treats. Some parts of the original of the following treatise, it must be acknowledged, are so confused, and some so minute, that it is extremely difficult to give an exact interpretation. But the general tenor, though not elegant, is plain and perspicuous; and such has it been here retained.

LETTER I.

ON THE IMPREGNATION OF THE QUEEN BEE.

SIR,

When I had the honour at Genthod of giving you an account of my principal experiments on bees, you desired me to transmit a written detail, that you might consider them with greater attention. I hasten, therefore, to extract the following observations from my journal.—As nothing can be more flattering to me than the interest you take in my researches, permit me to remind you of your promise to suggest new experiments[*].

After having long studied bees in glass hives constructed on M. de Reaumur's principle, you have found the form unfavourable to an observer. The hives being too wide, two parallel combs were made by the bees, consequently whatever passed between them escaped observation. From this inconvenience, which I have experienced, you recommended much thinner hives to naturalists, where the panes should be so near each other, that only a single row of combs could be erected between them. I have followed your admonitions, Sir, and provided hives only eighteen lines in width, in which I have found no difficulty to establish swarms. However, bees must not be entrusted with the charge of constructing a single comb:

---

[*] All these letters are addressed to the celebrated naturalist M. Bonnet.—*T.*

Nature has taught them to make parallel ones, which is a law they never derogate from, unless when constrained by some particular arrangement. Therefore, if left to themselves in these thin hives, as they cannot form two combs parallel to the plane of the hive, they will form several small ones perpendicular to it, and, in that case, all is equally lost to the observer. Thus it became essential previously to arrange the position of the combs. I forced the bees to build them perpendicular to the horizon, and so that the lateral surfaces were three or four lines from the panes of the hive. This distance allows the bees sufficient liberty, but prevents them from collecting in too large clusters on the surface of the comb. By such precautions, bees are easily established in very thin hives. There they pursue their labours with the same assiduity and regularity; and, every cell being exposed, none of their motions can be concealed.

It is true, that by compelling these insects to a habitation where they could construct only a single row of combs, I had, in a certain measure, changed their natural situation, and this circumstance might possibly have affected their instinct. Therefore, to obviate every objection, I invented a kind of hives, which, without losing the advantages of those very thin, at the same time approached the figure of common hives where bees form several rows of combs.

I took several small fir boxes, a foot square and fifteen lines wide, and joined them together by hinges, so that they could be opened and shut like the leaves of a book[*]. When using a hive of

---

[*] The leaf or book hive consists of twelve vertical frames or boxes, parallel to each other, and joined together. Fig. 1. the sides, f f. f g. should be twelve inches long, and the cross spars, f f. g g. nine or ten; the thickness of these spars an inch, and their breadth fifteen lines. It is necessary that this last measure should be accurate; a a. a piece of comb which guides the bees in their work; d. a moveable slider supporting the lower part; b b. pegs to keep the comb properly in the frame or box; four are in the opposite side; e e. pegs in the sides under the moveable slider to support it.

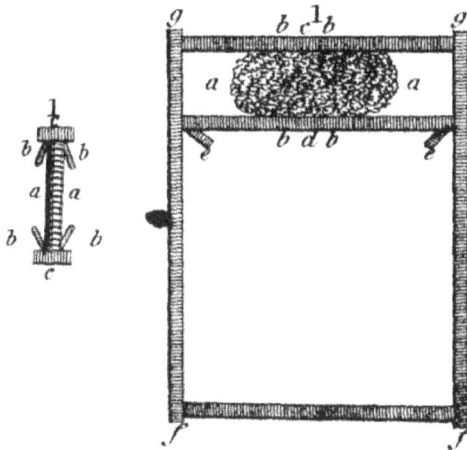

Figure 1

A book hive, consisting of twelve frames, all numbered, is represented fig. 2. Between 6 and 7 are two cases with lids, that divide the hive into two equal parts, and should only be used to separate the bees for forming an artificial swarm; a a. two frames which shut up the two sides of the hive, have sliders, b. b.

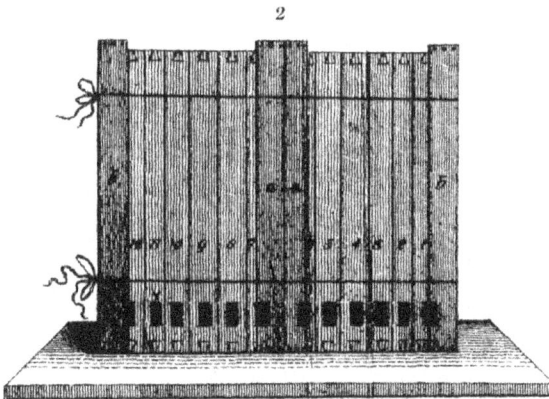

Figure 2

The entrance appears at the bottom of each frame. All should be close but 1 and 12. However it is necessary that they should open at pleasure.

The hive is partly open, fig. 3. and shews how the component parts may be united by hinges, and open as the leaves of a book. The two covers closing up the sides, a. a.

this description, we took care to fix a comb in each frame, and then introduced all the bees necessary for each particular experiment. By opening the different divisions successively, we daily inspected both surfaces of every comb. There was not a single cell where we could not distinctly see what passed at all times, nor a single bee, I may almost say, with which we were not particularly acquainted. Indeed, this construction is nothing more than the union of several very flat hives which may be separated. Bees, in such habitations, must not be visited before their combs are securely fixed in the frames, otherwise, by falling out, they may kill or hurt them, as also irritate them to that degree that the observer cannot escape stinging, which is always painful, and sometimes dangerous: but they soon become accustomed to their situation, and in some measure tamed by it; and, in three days, we may begin to operate on the hive, to open it, remove part of the combs, and substitute others, without the bees exhibiting too formidable symptoms of displeasure. You will remember, Sir, that on visiting my retreat, I shewed you a hive of this kind that had been a long time in experiment, and

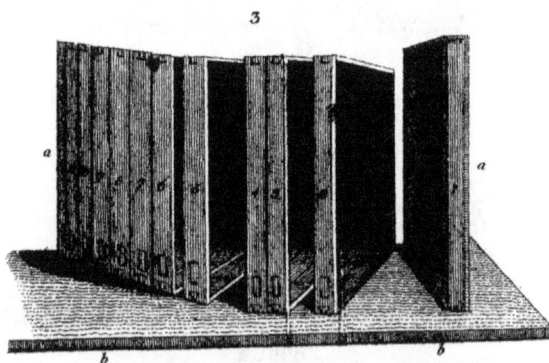

Figure 3

Fig. 4. [missing from original - Ed.] is another view of fig. 1. a a. a piece of comb to guide the bees; b b. pegs disposed so as to retain the comb properly in the frame; c c. parts of two shelves; the one above is fixed, and keeps the comb in a vertical position; the under one, which is moveable, supports it below.

how much you were surprised that the bees so quietly allowed us to open it.

In these hives, I have repeated all my observations, and obtained exactly the same results as in the thinnest. Thus, I think, already to have obviated any objections that may arise concerning the supposed inconvenience of flat hives. Besides, I cannot regret the repetition of my labours; by going over the same course several times, I am much more certain of having avoided error; and it also appears, that some advantages are found in these which may be called *Book* or *Leaf-hives*, as they prove extremely useful in the economical treatment of bees, which shall afterwards be detailed.

I now come to the particular object of this letter, the fecundation of the queen bee; and I shall, in a few words, examine the different opinions of naturalists on this singular problem. Next I shall state the most remarkable observations which their conjectures have induced me to make, and then describe the new experiments by which I think I have solved the problem[*].

Swammerdam, who studied bees with unremitting attention, and who never could see a real copulation between a drone and a queen, was satisfied that copulation was unnecessary for fecundation of the eggs: but having remarked that, at certain times, the drones exhaled a very strong odour, he thought this odour was an emanation of the *aura seminalis*, or the *aura seminalis* itself, which operated fecundation by penetrating the body of the female. His conjecture was confirmed on dissecting

---

[*] I cannot insist that my readers, the better to comprehend what is here said, shall peruse the Memoirs of M. de Reaumur on Bees, and those of the Lusaçe Society; but I must request them to examine the extracts in M. Bonnet's works, tom. 5. 4to edit. and tom. 10. 8vo, where they will find a short and distinct abstract of all that naturalists have hitherto discovered on the subject.

the male organs of generation; for he was so much struck with the disproportion between them and those of the female, that he did not believe copulation possible. His opinion, concerning the influence of the odour, had this farther advantage, that it afforded a good reason for the prodigious number of the males. There are frequently fifteen hundred or two thousand in a hive; and, according to Swammerdam, it is necessary they should be numerous, that the emanation proceeding from them may have an intensity or energy sufficient to effect impregnation.

Though M. de Reaumur has refuted this hypothesis by just and conclusive reasoning, he has failed to make the sole experiment that could support or overturn it. This was to confine all the drones of a hive in a tin case, perforated with minute holes, which might allow the emanation of the odour to escape, but prevent the organs of generation from passing through. Then, this case should have been placed in a hive well inhabited, but completely deprived of males, both of large and small size, and the consequence attended to. It is evident, had the queen laid eggs after matters were thus disposed, that Swammerdam's hypothesis would have acquired probability; and on the contrary it would have been confuted had she produced no eggs, or only sterile ones. However the experiment has been made by us, and the queen remained barren; therefore, it is undoubted, that the emanation of the odour of the males does not impregnate bees.

M. de Reaumur was of a different opinion. He thought that the queen's fecundation followed actual copulation. He confined several drones in a glass vessel along with a virgin queen: he saw the female make many advances to the males; but, unable to observe any union so intimate that it could be denominated copulation, he leaves the question undecided. We have repeated this experiment: we have frequently confined virgin queens with drones of all ages: we have done so at every season, and witnessed all their advances and solicitations to the males: we have even believed we saw a kind of union between them, but so short and imperfect that it was unlikely to effect impregnation. Yet, to neglect nothing, we confined the virgin queen, that had suffered the approaches of the male, to her hive. During a month that her imprisonment continued, she did not lay a single egg; therefore, these momentary junctions do not accomplish fecundation.

In the *Contemplation de la Nature*, you have cited the observations of the English naturalist Mr Debraw. They appear correct, and at last to elucidate the mystery. Favoured by chance, the observer one day perceived at the bottom of cells containing eggs, a whitish fluid, apparently spermatic, at least, very different from the substance or jelly which bees commonly collect around their new hatched worms. Solicitous to learn its origin, and conjecturing that it might be the male prolific fluid, he began to watch the motions of every drone in the hive, on purpose to seize the moment when they would bedew the eggs. He assures us, that he saw several insinuate the posterior part of the body into the cells, and there deposit the fluid. After frequent repetition of the first, he entered on a long series of experiments. He confined a number of workers in glass bells along with a queen and several males. They were supplied with pieces of comb containing honey, but no brood. He saw the queen lay eggs, which were bedewed by the males, and from which larvæ were hatched, consequently, he could not hesitate advancing as a fact demonstrated, that male bees fecundate the queen's eggs in the manner of frogs and fishes, that is, after they are produced.

There was something very specious in this explanation: the experiments on which it was founded seemed correct; and it afforded a satisfactory reason for the prodigious number of males in a hive. At the same time, the author had neglected to answer one strong objection. Larvæ appear when there are no drones. From the month of September until April, hives are generally destitute of males, yet, notwithstanding their absence, the queen then lays fertile eggs. Thus, the prolific fluid cannot be required to impregnate them, unless we can suppose that it is necessary at a certain time of the year, while at every other season it is useless.

To discover the truth amidst these facts apparently so contradictory, I wished to repeat Mr Debraw's experiments, and to observe more precaution than he himself had done. First, I sought for the fluid, which he supposes the seminal, in cells containing eggs. Several were actually found with that appearance; and, during the first days of observation, neither my assistant nor myself doubted the reality of the discovery. But we afterwards found it an illusion arising from the reflection of the light, for nothing like a fluid was visible, except when the solar rays reached the bottom of the cells. Fragments of the coccoons of worms, successively hatched, commonly cover the bottom; and, as they are shining, it may easily be conceived that, when much illuminated, an illusory effect results from the light. We proved it by the strictest examination, for no vestiges of a fluid were perceptible when the cells were detached and cut asunder.

Though the first observation inspired us with some distrust of Mr Debraw's discovery, we repeated his other experiments with the utmost care. On the 6. of August 1787, we immersed a hive, and, with scrupulous attention, examined the whole bees while in the bath. We ascertained that there was no male, either large or small; and having examined all the combs, we found neither male nymph, nor worm. When the bees were dry, we replaced them all, along with the queen, in their habitation, and transported them into my cabinet. They were allowed full liberty; therefore, they flew about, and made their usual collections; but, it being necessary that no male should enter the hive during the experiment, a glass tube was adapted to the entrance, of such dimensions that two bees only could pass at once; and we watched the tube attentively during the four or five days that the experiment continued. We should have instantly observed and removed any male that appeared, that the result of the experiment might be undisturbed, and I can positively affirm that not one was seen. However, from the first day, which was the sixth of August, the queen deposited fourteen eggs in the workers cells; and all these were hatched on the tenth of the same month.

This experiment is decisive, since the eggs laid by the queen of a hive where there were no males, and where it was impossible one could be introduced, since these eggs, I say, were fertile, it becomes indubitable that the fluid of the males is not required for their exclusion.

Though it did not appear that any reasonable objection could be started against this conclusion, yet, as I had been accustomed in all my experiments to seek for the most trifling difficulties that could arise, I conceived that Mr Debraw's partisans might maintain, that the bees, deprived of drones, perhaps would search for those in other hives, and carry the fecundative fluid to their own habitations for depositing it on the eggs.

It was easy to appreciate the force of this objection, for all that was necessary was a repetition of the former experiments, and to confine the bees so closely to their hives that none could possibly escape. You very well know, Sir, that these animals can live three or four months confined in a hive well stored with honey and wax, and if apertures are left for circulation of the air. This experiment was made on the tenth of August; and I ascertained, by means of immersion, that no male was present. The bees were confined four days in the closest manner, and then I found forty young larvæ.

I extended the precautions so far as to immerse this hive a second time, to assure myself that no male had escaped my researches. Each of the bees was separately examined, and none was there that did not display its sting. The coincidence of this experiment with the other, proved that the eggs were not externally fecundated.

In terminating the confutation of Mr Debraw's opinion, I have only to explain what led him into error; and that was, his using queens whose history he was unacquainted with from their origin. When he observed the eggs produced by a queen, confined along with males, were fertile, he thence concluded that they had been bedewed by the prolific fluid in the cells: but to render his conclusion just, he should first have ascertained that the female never had copulated, and this he neglected. The truth is, that, without knowing it, he had used, in his experiments, a queen after she had commerce with the male. Had he taken a virgin queen the moment she came from the royal cell, and confined her along with drones in his vessels, the result would have been opposite; for, even amidst a seraglio of males, this young queen would never have laid, as I shall afterwards prove.

The Lusatian observers, and M. Hattorf in particular, thought the queen was fecundated by herself, without concourse with the males. I shall here give an abstract of the experiment on which that opinion is founded.[*]

M. Hattorf took a queen whose virginity he could not doubt. He excluded all the males both of the large and small species, and, in several days, he found both eggs and worms. He asserts that there were no drones in the hive, during the course of the experiment; but although they were absent, the queen laid eggs, from which came worms: whence he considers she is impregnated by herself.

Reflecting on this experiment, I do not find it sufficiently accurate. Males pass with great facility from hive to hive; and

---

[*] Vide M. Schirach's History of Bees, in a memoir by M. Hattorf, entitled, *Physical Researches whether the Queen Bee requires fecundation by Drones?*

M. Hattorf took no precaution that none was introduced into his. He says, indeed, there was no male, but is silent respecting the means he adopted to prove the fact. Though he might be satisfied of no large drone being there, still a small one might have escaped his vigilance, and fecundated the queen. With a view to clear up the doubt, I resolved to repeat his experiment, in the manner described, and without greater care or precaution.

I put a virgin queen into a hive, from which all the males were excluded, but the bees left at perfect liberty. For several days I visited the hive, and found new hatched worms in it. Here then is the same result as M. Hattorf obtained? But before deducing the same consequence from it, we had to ascertain beyond dispute that no male had entered the hive. Thus, it was necessary to immerse the bees, and examine each separately. By this operation, we actually found four small males. Therefore, to render the experiment decisive, not only was it requisite to remove all the drones, but also, by some infallible method, to prevent any from being introduced, which the German naturalist had neglected.

I prepared to repair this omission, by putting a virgin queen into a hive, from which the whole males were carefully removed; and to be physically certain that none should enter, a glass tube was adapted at the entrance of such dimensions that the working bees could freely pass and repass, but too narrow for the smallest male. Matters continued thus for thirty days, the workers departing and returning performed their usual labours: but the queen remained sterile. At the expiration of this time, her belly was equally slender as at the moment of her origin. I repeated the experiment several times, and always with the same consequence.

Therefore, as a queen, rigorously separated from all commerce with the male, remains sterile, it is evident she cannot impregnate herself, and M. Hattorf's opinion is ill-founded.

Hitherto, by endeavouring to confute or verify the conjectures of all the authors who had preceded me, by new experiments, I acquired the knowledge of new facts, but these were apparently so contradictory as to render the solution of the problem still more difficult. While examining Mr. Debraw's hypothesis, I confined a queen in a hive, from which all the drones were removed; the queen nevertheless was fertile. When considering the opinion of M. Hattorf on the contrary, I put a queen, of whose virginity I was perfectly satisfied, in the same situation, she remained sterile.

Embarrassed by so many difficulties, I was on the point of abandoning the subject of my researches, when at length by more attentive reflection, I thought these contradictions might arise from experiments made indifferently on virgin queens, and on those with whose history I was not acquainted from the origin, and which had perhaps been impregnated unknown to me. Impressed with this idea, I undertook a new method of observation not on queens fortuitously taken from the hive, but on females decidedly in a virgin state, and whose history I knew from the instant they left the cell.

From a very great number of hives, I removed all the virgin females, and substituted for each a queen taken at the moment of her birth. The hives were then divided into two classes. From the first, I took the whole males both large and small, and adapted a glass tube at the entrance, so narrow, that no drone could pass, but large enough for the free passage of the common bees. In the hives of the second class, I left all the drones belonging to them, and even introduced more; and to prevent them from escaping, a glass tube, also too narrow for the males, was adapted to the entrance of these hives.

For more than a month, I carefully watched this experiment, made on a large scale; but much to my surprise, all the queens remained sterile. Thus it was proved, that queens confined in a hive would continue barren though amidst a seraglio of males.

This result induced me to suspect that the females could not be fecundated in the interior of the hive, and that it was necessary for them to leave it for receiving the approaches of the male. To ascertain the fact was easy, by a direct experiment; and as the point is important, I shall relate in detail what was done by my secretary and myself on the 29. June 1788.

Aware, that in summer the males usually leave the hive at the warmest time of the day, it was natural for me to conclude that if the queens were also obliged to go out for impregnation, instinct would induce them to do so at the same time as the males.

At eleven in the forenoon, we placed ourselves opposite a hive containing an unimpregnated queen five days old. The sun had shone from his rising; the air was very warm; and the males began to leave the hives. We then enlarged the entrance of that which we wished to observe, and paid great attention to the bees that entered and departed. The males appeared, and immediately took flight. Soon afterwards, the young queen appeared at the entrance; at first she did not fly, but brushed her belly with her hind legs, and traversed the board a little; neither workers nor males paid any attention to her. At last, she took flight. When several feet from the hive, she returned, and approached it as if to examine the place of her departure, perhaps judging this precaution necessary to recognize it; she then flew away, describing horizontal circles twelve or fifteen feet above the earth. We contracted the entrance of the hive that she might not return unobserved, and placed ourselves in the centre of the circles described in her flight, the more easily to follow her and observe all her motions. But she did not remain long in a situation favourable for us, and rapidly rose out of sight. We resumed our place before the hive; and in seven minutes, the

15

young queen returned to the entrance of a habitation which she had left for the first time. Having found no external appearance of fecundation, we allowed her to enter. In a quarter of an hour she re-appeared; and, after brushing herself as before, took flight. Then returning to examine the hive, she rose so high that we soon lost sight of her. Her second absence was much longer than the first; twenty-seven minutes elapsed before she came back. We then found her in a state very different from that in which she was after her first excursion. The sexual organs were distended by a white substance, thick and hard, very much resembling the fluid in the vessels of the male, completely similar to it indeed in colour and consistence[*].

But more evidence than mere resemblance was requisite to establish that the female had returned with the prolific fluid of the males. We allowed this queen to enter the hive, and confined her there. In two days, we found her belly swoln; and she had already laid near an hundred eggs in the worker's cells.

To confirm our discovery, we made several other experiments, and with the same success. I shall continue to transcribe my journal.

On the second of July, the weather being very fine, numbers of males left the hives. We set at liberty an unimpregnated

---

[*] It will afterwards appear that what we took for the generative fluid, was the male organs of generation, left by copulation in the body of the female. This discovery we owe to a circumstance that shall immediately be related. Perhaps I should avoid prolixity, by suppressing all my first observations on the impregnation of the queen, and by passing directly to the experiments that prove she carries away the genital organs; but in such observations which are both new and delicate, and where it is so easy to be deceived, I think service is done to the reader by a candid avowal of my errors. This is an additional proof to so many others, of the absolute necessity that an observer should repeat all his experiments a thousand times, to obtain the certainty of seeing facts as they really exist.

young queen, eleven days old, whose hive had always been deprived of males. Having quickly left the hive, she returned to examine it, and then rose out of sight. In a few minutes, she returned without any external marks of impregnation. In a quarter of an hour, she departed again, but her flight was so rapid that we could scarcely follow her a moment. This absence continued thirty minutes. On returning, the last ring of the body was open, and the sexual organs full of the whitish substance already mentioned. She was then replaced in the hive from which all the males were excluded. In two days, we found her impregnated.

These observations at length demonstrate why M. Hattorf obtained results so different from ours. His queens, though in hives deprived of males, had been fecundated, and he thence concludes that sexual intercourse is not requisite for their impregnation. But he did not confine the queens to their hives, and they had profited by their liberty to unite with the males. We, on the contrary, have surrounded our queens with a number of males; but they continued sterile; because the precaution of confining the males to their hives had also prevented the queens from departing to seek that fecundation without, which they could not obtain within.

These experiments were repeated on queens, twenty, twenty-five, and thirty days old. All became fertile after a single impregnation; however, we have remarked some essential peculiarities in the fecundity of those unimpregnated until the twentieth day of their existence; but we shall defer speaking of the fact until we can present naturalists with observations sufficiently secure and numerous to merit their attention: Yet let me add a few words more. Though neither my assistant nor myself have witnessed the copulation of a queen and a drone, we think that, after the detail which has just been commenced, no doubt of it can remain, or of the necessity of copulation to effect impregnation. The sequel of experiments, made with every possible precaution, appears demonstrative. The uniform sterility of queens in hives wanting males, and in those where they were confined along with them; the departure of these queens from the hives; and the very conspicuous evidence of impregnation with which they return, are proofs against which no objections can stand. But we do not despair of being able next spring to obtain the complement of this proof, by seizing the female at the very moment of copulation.

Naturalists have always been very much embarrassed to account for the number of males found in most hives, and which seem only a burden on the community, since they fulfil no function. But we now begin to discern the object of nature in multiplying them to that extent. As fecundation cannot be accomplished within, and as the queen is obliged to traverse the expanse of the atmosphere, it is requisite the males should be numerous that she may have the chance of meeting some one of them. Were only two or three drones in each hive, there would be little probability of their departure at the same instant with the queen, or that they would meet in their excursions; and most of the females would thus remain sterile.

But why has nature prohibited copulation within the hives? This is a secret still unknown to us. It is possible, however, that some favourable circumstance may enable us to penetrate it in the course of our observations. Various conjectures may be formed; but at this day we require facts, and reject gratuitous suppositions. It should be remembered, that bees do not form the sole republic among insects presenting a similar phenomenon; female ants are also obliged to leave the ant-hills previous to fecundation.

I cannot request, Sir, that you will communicate the reflections which your genius will excite concerning the facts I have related. This is a favour to which I am not yet entitled. But as new experiments will unquestionably occur to you, whether on the impregnation of the queen or on other points, may I solicit you to suggest them? They shall be executed with all possible care; and I shall esteem this mark of friendship and interest as the most flattering encouragement that the continuance of my labours can receive.

PREGNY, 13th August 1789.

◆　◆　◆　◆　◆

LETTER FROM M. BONNET TO M. HUBER.

You have most agreeably surprised me, Sir, with your interesting discovery of the impregnation of the queen bee. It was a fortunate idea, that she left the hive to be fecundated, and your method of ascertaining the fact was extremely judicious and well adapted to the object in view.

Let me remind you, that male and female ants copulate in the air; and that after impregnation the females return to the ant hills to deposit their eggs. *Contemplation de la Nature, Part II. chap. 22. note 1.* It would be necessary to seize the instant when the drone unites with the female. But how remote from the power of the observer are the means of ascertaining a copulation in the air. If you have satisfactory evidence that the fluid bedewing the last rings of the female is the same with that of the male, it is more than mere presumption in favour of copulation. Perhaps it may be necessary that the male should seize the female under the belly, which cannot easily be done but in the air. The large opening at the extremity of the queen, which you have observed in so particular a condition, seems to correspond to the singular size of the sexual parts of the male.

You wish, my dear Sir, that I should suggest some new experiments on these industrious republicans. In doing so, I shall take the greater pleasure and interest, as I know to what extent you possess the valuable art of combining ideas, and of deducing from this combination results adapted to the discovery of new facts. A few at this moment occur to me.

It may be proper to attempt the artificial fecundation of a virgin queen, by introducing a little of the male's prolific fluid with a pencil, and at the same time observing every precaution to avoid error. Artificial fecundation, you are aware, has already succeeded in more than one animal.

To ascertain that the queen, which has left the hive for impregnation, is the same that returns to deposit her eggs, you will find it necessary to paint the thorax with some varnish that resists humidity. It will also be right to paint the thorax of a considerable number of workers in order to discover the duration of their life. This is a more secure method than slight mutilations.

For hatching the worm, the egg must be fixed almost vertically by one end near the bottom of the cell. Is it true, that it is unproductive unless fixed in this manner? I cannot determine the fact; and therefore leave it to the decision of experiment.

I formerly mentioned to you that I had long doubted the real nature of the small ovular substances deposited by queens in the cells, and my inclination to suppose them minute worms not yet begun to expand. Their elongated figure seems to favour my suspicions. It would therefore be proper to watch them with the utmost assiduity, from the instant of production until the period of exclusion. If the integument bursts, there can be no doubt that these minute substances are real eggs.

I return to the mode of operating copulation. The height that the queen and the males rise to in the air prevent us from seeing what passes between them. On that account, the hive should be put into an apartment with a very lofty ceiling. M. de Reaumur's experiment of confining a queen with several males in a glass vessel, merits repetition; and if, instead of a vessel, a glass tube, some inches in diameter and several feet long, were used, perhaps something satisfactory might be discovered.

You have had the fortune to observe the small queens mentioned by the Abbé Needham, but which he never saw. It will be of great importance to dissect them for the purpose of finding their ovaries. When M. Reims informed me that he had confined three hundred workers, along with a comb containing no eggs, and afterwards found hundreds in it, I strongly recommended that he should dissect the workers. He did so; and informed me that eggs were found in three. Probably without being aware of it, he has dissected small queens. As small drones exist, it is not surprising if small queens are produced also, and undoubtedly by the same external causes.

It is of much consequence to be intimately acquainted with this species of queens, for they may have great influence on different experiments and embarrass the observer: we should ascertain whether they inhabit pyramidal cells smaller than the common, or hexagonal ones.

M. Schirach's famous experiment on the supposed conversion of a common worm into a royal one, cannot be too often repeated, though the Lusatian observers have already done it frequently. I could wish to learn whether, as the discoverer maintains, the experiment will succeed only with worms, three or four days old, and never with simple eggs.

The Lusatian observers, and those of the Palatinate, affirm, that when common bees are confined with combs absolutely void of eggs, they then lay none but the eggs of drones. Thus, there must be small queens producing the eggs of males only, for it is evident they must have produced those supposed to come from workers. But how is it possible to conceive that their ovaries contain male eggs alone?

According to M. de Reaumur, the life of chrysalids may be prolonged by keeping them in a cold situation, such as an ice-house. The same experiment should be made on the eggs of a queen; on the nymphs of drones and workers.

Another interesting experiment would be to take away all the combs composing the common cells, and leave none but those destined for the larvæ of males. By this means we should learn whether the eggs of common worms, laid by the queen in the large cells, will produce large workers. It is very probable, however, that deprivation of the common cells might discourage the bees, because they require them for their honey and wax. Nevertheless, it is likely, by taking away only part of the common cells, the workers may be forced to lay common eggs in the cells of drones.

I should also wish to have the young larvæ gently removed from the royal cell, and deposited at the bottom of a common one, along with some of the royal food.

As the figure of hives has much influence on the respective disposition of the combs, it would be a satisfactory experiment, greatly to diversify their shape and internal dimensions. Nothing could be better adopted to instruct us how bees can regulate their labours, and apply them to existing circumstances. This may enable us to discover particular facts which we cannot foresee.

The royal eggs and those producing drones, have not yet been carefully compared with the eggs from which workers come. But they ought to be so, that we may ascertain whether these different eggs have secret distinctive characteristics.

The food supplied by the workers to the royal worm, is not the same with that given to the common worm. Could we not endeavour, with the point of a pencil, to remove a little of the royal food, and give it to a common worm deposited in a cell of the largest dimensions? I have seen common cells hanging almost vertically, where the queen had laid; and these I should prefer for this experiment.

Various facts, which require corroboration, were collected in my Memoirs on Bees; of this number are my own observations. You can select what is proper, my dear Sir. You have already enriched the history of bees so much, that every thing may be expected from your understanding and perseverance. You know the sentiments with which you have inspired the CONTEMPLATOR OF NATURE.

GENTHOD, 18. August 1789.

LETTER II.

SEQUEL OF OBSERVATIONS ON THE IMPREGNATION OF THE QUEEN BEE.

SIR,

All the experiments, related in my preceding letter, were made in 1787 and 1788. They seem to establish two facts, which had previously been the subject of vague conjecture: 1. The queen bee is not impregnated of herself, but is fecundated by copulation with the male. 2. Copulation is accomplished without the hive, and in the air.

The latter appeared so extraordinary, that notwithstanding all the evidence obtained of it, we eagerly desired to take the queen in the fact; but, as she always rises to a great height, we never could see what passed. On that account you advised us to cut part off the wings of virgin queens. We endeavoured to benefit by your advice, in every possible manner; but to our great regret, when the wings lost much, the bees could no longer fly; and, by cutting off only an inconsiderable portion, we did not diminish the rapidity of their flight. Probably there is a medium, but we were unable to attain it. On your suggestion, we tried to render their vision less acute, by covering the eyes with an opaque varnish, which was an experiment equally fruitless.

We likewise attempted artificial fecundation, and took every possible precaution to insure success. Yet the result was always unsatisfactory. Several queens were the victims of our curiosity; and those surviving remained sterile. Though these different experiments were unsuccessful, it was proved that queens leave their hives to seek the males, and that they return with undoubted evidence of fecundation. Satisfied with this, we could only trust to time or accident for decisive proof of an actual copulation. We were far from suspecting a most singular discovery, which we made in July this year, and which affords complete demonstration of the supposed event, namely, that the sexual organs of the male remain with the female.[*]

---

[*] The remainder of this Chapter chiefly consists of anatomical details. These may rather be considered an interruption of the narrative; and the Translator has judged it expedient to transfer them to an Appendix.

LETTER III.

THE SAME SUBJECT CONTINUED.—OBSERVATIONS ON RETARDING THE FECUNDATION OF QUEENS.

In my first letter, I remarked, that when queens were prevented from receiving the approaches of the male until the twenty-fifth or thirtieth day of their existence, the result presented very interesting peculiarities. My experiments at that time were not sufficiently numerous; but they have since been so often repeated, and the result so uniform, that I no longer hesitate to announce, as a certain discovery, the singularities which retarded fecundation, produces on the ovaries of the queen. If she receives the male during the first fifteen days of her life, she remains capable of laying both the eggs of workers and of drones; but should fecundation be retarded until the twenty-second day, her ovaries are vitiated in such a manner that she becomes unfit for laying the eggs of workers, and will produce only those of drones.

In June 1787, being occupied in researches relative to the formation of swarms, I had occasion, for the first time, to observe a queen that laid none but the eggs of males. When a hive is ready to swarm, I had before observed, that the moment of swarming is always preceded by a very lively agitation, which first affects the queen, is then communicated to the workers, and excites such a tumult among them, that they abandon their labours, and rush in disorder to the outlets of the hive. I then knew very well the cause of the queen's agitation, and it is described in the history of swarms, but I was ignorant how the delirium communicated to the workers; and this difficulty interrupted my researches. I therefore thought of investigating, by direct experiments, whether at all times, when the queen was greatly agitated, even not in the time of the hive swarming, her agitation would in like manner be communicated to the workers. The moment a queen was hatched, I confined her to the hive by contracting the entrances. When assailed by the imperious desire of union with the males, I could not doubt that she would make great exertions to escape, and that the impossibility of it would produce a kind of delirium. I had the patience to observe this queen thirty-four days. Every morning about eleven o'clock, when the weather was fine and the sunshine invited the males to leave their hives, I saw her impetuously traverse every corner of her habitation, seeking to escape. Her fruitless efforts threw her into an uncommon agitation, the symptoms of which I shall elsewhere describe, and all the common bees were affected by it. As she never was out all this time, she could not be impregnated. At length, on the thirty-sixth day, I set her at liberty. She soon took advantage of it; and was not long of returning with the most evident marks of fecundation.

Satisfied with the particular object of this experiment, I was far from any hopes that it would lead to the knowledge of another very remarkable fact; how great was my astonishment, therefore, on finding that this female, which, as usual, began to lay forty-six hours after copulation, laid the eggs of drones, but none of workers, and that she continued ever afterwards to lay those of drones only.

At first, I exhausted myself with conjectures on this singular fact; the more I reflected on it, the more did it seem inexplicable. At length, by attentively meditating on the circumstances of the experiment it appeared there were two principles, the influence of which I should first of all endeavour to appreciate separately. On the one hand, this queen had suffered long confinement; on the other, her fecundation had been extremely retarded. You know, Sir, that queens generally receive the males about the fifth or sixth day, and this queen had not copulated until the thirty-sixth. Little weight could be given to the supposition, that the peculiarity could be occasioned by confinement. Queens, in the natural state, leave their hives only once to seek the males. All the rest of their life they remain voluntary prisoners. Thus, it was improbable that captivity could produce the effect I wished to explain. At the same time, as it was essential to neglect nothing in a subject so new, I wished to ascertain whether it was owing to the length of confinement, or to retarded fecundation.

Investigating this was no easy matter. To discover whether captivity, and not retarded fecundation, vitiated the ovaries, it was necessary to allow a female to receive the approaches of a male, and also to keep her imprisoned. Now this could not be, for bees never copulate in hives. On the same account, it was impossible to retard the copulation of a queen without keeping her in confinement. I was long embarrassed by the difficulty. At length, I contrived an apparatus, which, though imperfect, nearly fulfilled my purpose.

I put a queen, at the moment of her last metamorphosis, into a hive well stored, and sufficiently provided with workers and males; the entrance was contracted so as to prevent her exit, but allowed free passage to the workers. I also made another opening for the queen, and adapted a glass tube to it, communicating with a cubical glass box eight feet high. Hither the queen could at all times come and fly about, enjoying a purer air than was to be found within the hive; but she could not be fecundated; for though the males flew about within the same bounds, the space was too limited to admit of any union between them. By the experiments related in my first letter, copulation takes place high in the air only: therefore, in this apparatus, I found the advantage of retarding fecundation, while the liberty the queen now had, did not render her situation too remote from the natural state. I attended to the experiment fifteen days. Every fine morning, the young captive left her hive; she traversed her glass prison, and flew much about, and with great facility. She laid none during this interval, for she had not united with a male. On the sixteenth day, I set her at liberty: she left the hive, rose aloft in the air, and soon returned with full evidence of impregnation. In two days, she laid, first the eggs of workers, and afterwards as many as the most fertile queens.

It thence followed, 1. That captivity did not alter the organs of queens. 2. When fecundation took place within the first sixteen days, she produced both species of eggs.

This was an important experiment. It rendered my labours much more simple, by clearly pointing out the method to be pursued: it absolutely precluded the supposed influence of captivity; and left nothing for investigation but the consequences of retarded fecundation.

With this view, I repeated the experiment; but, instead of giving the virgin queen liberty on the sixteenth day, I retained her until the twenty-first. She departed, rose high in the air, was fecundated, and returned. Thirty-six hours afterwards, she began to lay: but it was the eggs of males only, and, although very fruitful afterwards, she laid no other kind.

I occupied myself the remainder of 1787, and the two subsequent years, with experiments on retarded fecundation, and had constantly the same results. It is undoubted, therefore, that when the copulation of queens is retarded beyond the twentieth day, only an imperfect impregnation is operated: instead of laying the eggs of workers and males equally, they will lay none but those of males.

I do not aspire to the honour of explaining this singular fact. When the course of my experiments led me to observe that some queens laid only the eggs of drones, it was natural to investigate the proximate cause of such a singularity; and I ascertained that it arose from retarded fecundation. My evidence is demonstrative, for I can always prevent queens from laying the eggs of workers, by retarding their fecundation until the twenty-second or twenty-third day. But, what is the remote cause of this peculiarity; or, in other words, why does the delay of impregnation render queens incapable of laying the eggs of workers? This is a problem on which analogy throws no light: nor in all physiology am I acquainted with any fact that bears the smallest similarity.

The problem becomes still more difficult by reflecting on the natural state of things, that is when fecundation has not been delayed. The queen then lays the eggs of workers forty-six hours after copulation, and continues for the subsequent eleven months to lay these alone: and it is only after this period that a considerable and uninterrupted laying of the eggs of drones commences. When, on the contrary, impregnation is retarded after the twentieth day, the queen begins, from the forty-sixth hour, to lay the eggs of males, and no other kind during her whole life. As, in the natural state, she lays the eggs of workers only, during the first eleven months, it is clear that these, and the male eggs, are not indiscriminately mixed in the oviducts. Undoubtedly they occupy a situation corresponding to the principles that regulate laying: the eggs of workers are first, and those of drones behind them. Farther, it appears that the queen can lay no male eggs until those of workers, occupying the first place in the oviducts, are discharged. Why, then, is this order inverted by retarded copulation? How does it happen that all the workers eggs which the queen ought to lay, if fecundation was in due time, now wither and disappear, yet do not, impede the passage of the eggs of drones, which occupy only the second place in the ovaries. Nor is this all. I have satisfied myself that a single copulation is sufficient to impregnate the whole eggs that a queen will lay in the course of at least two years. I have even reason to think, that a single copulation will impregnate all the eggs that she will lay during her whole life: but I want absolute proof for more than two years. This, which is truly a very singular fact in itself, renders the influence of retarded fecundation still more difficult to be accounted for. Since a single copulation suffices, it is clear that the male fluid acts from the first moment on all the eggs that the queen will lay in two years. It gives them, according to your principles, that degree of *animation* that afterwards effects their successive expansion. Having received the first impressions of life, they grow, they mature, so to speak, until the day they are laid: and as the laws of laying are constant, because the eggs of the first eleven months are always those of workers, it is evident that those which appear first are also the eggs that come soonest to maturity. Thus, in the natural state, the space of eleven months is necessary for the male eggs to acquire that degree of increment they must have attained when laid. This consequence, which to me seems immediate, renders the problem insoluble. How can the eggs, which should grow slowly for eleven months, suddenly acquire their full expansion in forty-eight hours, when fecundation has been retarded twenty-one days, and by the effect of this retardation alone? Observe, I beseech you, that the hypothesis of successive expansion is not gratuitous; it rests on the principles of sound philosophy. Besides, for conviction that it is well founded, we have only to look at the figures given by Swammerdam of the ovaries of the queen bee. There we see eggs in that part of the oviducts contiguous to the vulva, much farther advanced, and larger than those contained in the opposite part. Therefore the difficulty remains in full force: it is an abyss where I am lost.

The only known fact bearing any relation to that now described, is the state of certain vegetable seeds, which, although extremely well preserved, lose the faculty of germination from age. The eggs of workers may also preserve, only for a very short time, the property of being fecundated by the seminal fluid;

and, after this period, which is about fifteen or eighteen days, become disorganised to that degree, that they can no longer be animated by it. I am sensible that the comparison is very imperfect; besides, it explains nothing, nor does it even put us on the way of making any new experiments. I shall add but one reflection more.

Hitherto no other effect has been observed from the retarded impregnation of animals, but that of rendering them absolutely sterile. The first instance of a female still preserving the faculty of engendering males, is presented by the queen bee. But as no fact in nature is unique, it is most probable that the same peculiarity will also be found in other animals. An extremely curious object of research would be to consider insects in this new point of view, I say *insects*, for I do not conceive that any thing analogous will be found in other species of animals. The experiments now suggested would necessarily begin with insects the most analogous to bees; as wasps, humble bees, mason bees, all species of flies, and the like. Some experiments might also be made on butterflies; and, perhaps, an animal might be found whose retarded fecundation would be attended with the same effects as that of queen bees. Should the animal be larger, dissection will be more easily accomplished; and we may discover what happens to the eggs when retarded fecundation prevents their expansion. At least, we might hope that some fortunate circumstance would lead to solution of the problem[*].

Let us now return to my experiments. In May 1789, I took two queens just when they had undergone the last metamorphosis: one was put in a *leaf hive*, well provided with

---

[*] The experiments suggested in this paragraph, recall a singular reflection of M. de Reaumur. Where treating of oviparous flies, he says, it would not be impossible for a hen to produce a living chicken, if, after fecundation, the eggs she should first lay could by any means be retained twenty-one days in the oviducts. *Mem. sur. les Insect. tom. 4. mem. 10.*

honey and wax, and sufficiently inhabited by workers and males. The other was put into a hive exactly similar, from which all the drones were removed. The entrances of these hives were too confined for the passage of the females and drones, but the common bees enjoyed perfect liberty. The queens were imprisoned thirty days; and being then set at liberty, they departed, and returned impregnated. Visiting the hives in the beginning of July, I found much brood, but wholly consisting of the worms and nymphs of males. There actually was not a single worker's worm or nymph. Both queens laid uninterruptedly until autumn, and constantly the eggs of drones. Their laying ended in the first week of November, as that of my other queens.

I was very earnest to learn what would become of them in the subsequent spring, whether they would resume laying, or if new fecundation would be necessary; and if they did lay, of what species the eggs would be. However, the hives being very weak, I dreaded they might perish during winter. Fortunately, we were able to preserve them; and from April 1790, they recommenced laying. The precautions we had taken prevented them from receiving any new approaches of the male. Their eggs were still those of males.

It would have been extremely interesting to have followed the history of these two females still farther, but, to my great regret, the workers abandoned their hives on the fourth of May, and that same day I found both queens dead. No weevils were in the hive, which could disturb the bees; and the honey was still very plentiful: but as no workers had been produced in the course of the preceding year, and winter had destroyed many, they were too few in spring to engage in their wonted labours, and, from discouragement, deserted their habitation to occupy the neighbouring hives.

In my Journal, I find a detail of many experiments on the retarded impregnation of queen bees, so many, that transcribing the whole would be tedious. I may repeat, however, that there was not the least variation in the principle, and that whenever the copulation of queens was postponed beyond the twenty-first day, the eggs of males only were produced. Therefore, I shall limit my narrative to those experiments that have taught me some remarkable facts.

A queen being hatched on the fourth of October 1789, we put her into a leaf-hive. Though the season was well advanced, a considerable number of males was still in the hive; and it here became important to learn, whether, at this period of the year, they could equally effect fecundation; also, in case it succeeded, whether a laying, begun in the middle of autumn, would be interrupted or continued during winter. Thus, we allowed the queen to leave the hive. She departed, indeed, but made four and twenty fruitless attempts before returning with the evidence of fecundation. Finally, on the thirty-first of October, she was more fortunate: She departed, and returned with the most undoubted proof of the success of her amours: She was now twenty-seven days old, consequently fecundation had been retarded. She ought to have begun laying within forty-six hours, but the weather was cold, and she did not lay; which proves, as we may cursorily remark, that refrigeration of the atmosphere is the principal agent that suspends the laying of queens during winter. I was excessively impatient to learn whether, on the return of spring, she would prove fertile, without a new copulation. The means of ascertaining the fact was easy; for the entrances of the hives only required contraction, so as to prevent her from escaping. She was confined from the end of October until May. In the middle of March, we visited the combs, and found a considerable number of eggs, but, none being yet hatched, we could not know whether they would produce workers or males. On the fourth of April, having again examined the state of the hive, we found a prodigious quantity of nymphs and worms, all of drones; nor had this queen laid a single worker's egg.

Here, as well as in the preceding experiment, retardation had rendered the queens incapable of laying the eggs of workers. But this result is the more remarkable, as the queen did not commence laying until four months and a half after fecundation. It is not rigorously true, therefore, that the term of forty-six hours elapses between the copulation of the female and her laying; the interval may be much longer, if the weather grows cold. Lastly, it follows, that although cold will retard the laying of a queen impregnated in autumn, she will begin to lay in spring without requiring new copulation.

It may be added, that the fecundity of the queen, whose history is given here, was astonishing. On the first of May, we found in her hive, besides six hundred males, already flies, two thousand four hundred and thirty-eight cells, containing either eggs or nymphs of drones. Thus, she had laid more than three thousand male eggs during March and April, which is above fifty each day. Her death soon afterwards unfortunately interrupted my observation, I intended to calculate the total number of male eggs that she should lay throughout the year, and compare it with those of queens whose fecundation had not been retarded. You know, Sir, that the latter lay about two thousand male eggs in spring; and another laying, but less considerable, commences in August, also in the interval, that they produce the eggs of workers almost solely. But it is otherwise with the females whose copulation has been retarded: they produce no workers' eggs. For four or five months following, they lay the eggs of males without interruption, and in such numbers, that, in this short time, I suppose one queen gives birth to more drones than a female, whose fecundation has not been retarded, produces in the course of two years. It gives me much regret, that I have not been able to verify this conjecture.

I should also describe the very remarkable manner in which queens, that lay only the eggs of drones, sometimes deposit them in the cells. Instead of being placed in the lozenges forming the bottom, they are frequently deposited on the lower side of the cells, two lines from the mouth. This arises from the body of such queens being shorter than that of those whose fecundation has not been retarded. The extremity remains slender, while the first two rings next the thorax are uncommonly swoln. Thus, in disposing themselves for laying, the extremity cannot reach the bottom of the cells on account of the swoln rings; consequently the eggs must remain attached to the part that the extremity reaches. The worms proceeding from them pass their vermicular state in the same place where the eggs were deposited, which proves that bees are not charged with the care of transporting the eggs as has been supposed. But here they follow another plan. They extend beyond the surface of the comb those cells where they observe the eggs deposited, two lines from the mouth.

Permit me, Sir, to digress a moment from the subject, to give the result of an experiment which seems interesting. Bees, I say, are not charged with the care of transporting into cells, the eggs misplaced by the queen: and, judging by the single instance I have related, you will think me well entitled to deny this feature of their industry. However, as several authors have maintained the reverse, and even demanded our admiration of them in conveying the eggs, I should explain clearly that they are deceived.

I had a glass hive constructed of two stages; the higher was filled with combs of large cells, and the lower with those of common ones. A kind of division, or diaphraghm, separated these two stages from each other, having at each side an opening for the passage of the workers from one stage to the other, but too narrow for the queen. I put a considerable number of bees into this hive; and, in the upper part, confined a very fertile queen that had just finished her great laying of male eggs; therefore she had only those of workers to lay, and she was obliged to deposit them in the surrounding large cells from the want of others. My object in this arrangement will already be anticipated. My reasoning was simple. If the queen laid workers' eggs in the large cells, and the bees were charged with transporting them if misplaced, they would infallibly take advantage of the liberty allowed to pass from either stage: they would seek the eggs deposited in the large cells, and carry them down to the lower stage containing the cells adapted for that species. If, on the contrary, they left the common eggs in the large cells, I should obtain certain proof that they had not the charge of transporting them.

The result of this experiment excited my curiosity extremely. We observed the queen several days without intermission. During the first twenty-four hours, she persisted in not laying a single egg in the surrounding cells; she examined them one after another, but passed on without insinuating her belly into one. She was restless, and traversed the combs in all directions: her eggs appeared an oppressive burden, but she persisted in retaining them rather than they should be deposited in cells of unsuitable diameter. The bees, however, did not cease to pay her homage, and treat her as a mother. I was amused to observe, when she approached the edges of the division separating the two stages, that she gnawed at them to enlarge the passage: the workers approached her, and also laboured with their teeth, and made every exertion to enlarge the entrance to her prison, but ineffectually. On the second day, the queen could no longer retain her eggs: they escaped in spite of her, and fell at random. Then we conceived that the bees would convey them into the small cells of the lower stage, and we sought them there with the utmost assiduity; but I can safely affirm there was not one. The eggs that the queen still laid the third day disappeared as the first. We again sought them in the small cells, but none were there. The fact is, they are ate by the workers; and this is what has deceived the naturalists, who supposed them carried away. They have observed the misplaced eggs disappear, and, without farther investigation, have asserted that the bees convey them elsewhere: they take them, indeed, not to convey them any where, but to devour them. Thus nature has not charged bees with the care of placing the eggs in the cells appropriated for them, but she has inspired females themselves with sufficient instinct to know the species of eggs they are about to lay, and to deposit them in suitable cells. This has already been observed by M. de Reaumur, and here my observations correspond with his. Thus it is certain that in the natural state, when fecundation takes place at the proper time, and the queen has suffered from nothing, she is never deceived in the choice of the cells where her eggs are to be deposited; she never fails to lay those of workers in small cells, and those of males in large ones. The distinction is important, for the same certainty of instinct is no longer conspicuous in the conduct of those females whose impregnation has been deferred. I was oftener than once deceived respecting the eggs that such queens laid, for they were deposited indiscriminately in small cells and those of drones; and not aware of their instinct having suffered, I conceived that the eggs in small cells would produce workers; therefore I was very much surprised, when, at the moment they should have been hatched, the bees closed up the cells, and demonstrated, by anticipation, that the included worms would change into drones; they actually became males; those produced in small cells were small, those in large cells large. Thus I must warn observers, who would repeat my experiments on queens that lay only the eggs of males, not to be deceived by these circumstances, and expect that eggs of males will be deposited in the workers cells.

It is a singular fact, that the females, whose fecundation has been retarded, sometimes lay the eggs of males in royal cells. I shall prove, in the history of swarms, that immediately when queens, in the natural state, begin their great laying of male eggs, the workers construct numerous royal cells. Undoubtedly, there is some secret relation between the appearance of male eggs and the construction of these cells; for it is a law of nature from which bees never derogate. It is not surprising, therefore, that such cells are constructed in hives governed by queens laying the eggs of males only. It is no longer extraordinary that these queens deposit in the royal cells, eggs of the only species they can lay, for in general their instinct seems affected. But what I cannot comprehend is, why the bees take exactly the same care of the male eggs deposited in royal cells, as of those that should become queens. They provide them more plentifully with food, they build up the cells as if containing a royal worm; in a word, they labour with such regularity that we have frequently been deceived. More than once, in the firm persuasion of finding royal nymphs, we have opened the cells after they were sealed, yet the nymph of a drone always appeared. Here the instinct of the workers seemed defective. In the natural state, they can accurately distinguish the male worms from those of common bees, as they never fail giving a particular covering to the cells containing the former. Why then can they no longer distinguish the worms of drones when deposited in the royal cells? The fact deserves much attention. I am convinced that to investigate the instinct of animals, we must carefully observe where it appears to err.

Perhaps I should have begun this letter with an abstract of the observations of prior naturalists, on queens laying none but the eggs of males; however, I shall here repair the omission.

In a work, *Histoire de la Reine des Abeilles*, translated from the German by *Blassiere*, there is printed a letter from M. Schirach to you, dated 15 April 1771, where he speaks of some hives, in which the whole brood changed into drones. You will remember that he ascribes this circumstance to some unknown vice in the ovaries of the queen; but he was far from suspecting that retarded fecundation had been the cause of vitiation. He justly felicitated himself on discovering a method to prevent the destruction of hives in this situation, which was simple, for it consisted in removing the queen that laid the eggs of males only, and substituting one for her whose ovaries were not impaired. But to make the substitution effectual, it was necessary to procure queens at pleasure; a secret reserved for M. Schirach, and of which I shall speak in the following letter. You observe that the whole experiments of the German naturalist tended to the preservation of the hives whose queens laid none except male eggs; and that he did not attempt to discover the cause of the vice evident in their ovaries.

M. de Reaumur also says a few words, somewhere, of a hive containing many more drones than workers, but advances no conjectures on the cause. However, he adds, as a remarkable circumstance, that the males were tolerated in this hive until the subsequent spring. It is true that bees governed by a queen laying only male eggs, or by a virgin queen, preserve their drones several months after they have been massacred in other hives. I can ascribe no reason for it, but it is a fact I have several times witnessed during my long course of observations on retarded impregnation. In general it has appeared that while the queen lays male eggs, bees do not massacre the males already perfect in the hive.

PREGNY, 21. August 1791.

LETTER IV.

ON M. SCHIRACH'S DISCOVERY.

When you found it necessary, Sir, in the new edition of your works, to give an account of M. Schirach's beautiful experiments on the conversion of common worms into royal ones, you invited naturalists to repeat them. Indeed such an important discovery required the confirmation of several testimonies. For this reason, I hasten to inform you that all my researches establish the reality of the discovery. During ten years that I have studied bees, I have repeated M. Schirach's experiment so often, and with such uniform success, that I can no longer have the least doubt on the subject. Therefore, I consider it an established fact, when bees lose their queen, and several workers' worms are preserved in the hive, they enlarge some of their cells, and supply them not only with a different kind of food, but a greater quantity of it, and the worms reared in this manner, instead of changing to common bees, become real queens. I request my readers to reflect on the explanation you have given of so uncommon a fact, and the philosophical consequences you have deduced from it. *Contemplation de la Nature, part. II, chap. 27.*

In this letter I shall content myself with some account of the figure of the royal cells constructed by bees around those worms that are destined for the royal state, and terminate with discussing some points wherein my observations differ from those of M. Schirach.

Bees soon become sensible of having lost their queen, and in a few hours commence the labour necessary to repair their loss. First, they select the young common worms, which the requisite treatment is to convert into queens, and immediately begin with enlarging the cells where they are deposited. Their mode of proceeding is curious; and the better to illustrate it, I shall describe the labour bestowed on a single cell, which will apply to all the rest, containing worms destined for queens. Having chosen a worm, they sacrifice three of the contiguous cells: next, they supply it with food, and raise a cylindrical inclosure around, by which the cell becomes a perfect tube, with a rhomboidal bottom; for the parts forming the bottom are left untouched. If the bees damaged it, they would lay open three corresponding cells on the opposite surface of the comb, and, consequently, destroy their worms, which would be an unnecessary sacrifice, and Nature has opposed it. Therefore, leaving the bottom rhomboidal, they are satisfied with raising a cylindrical tube around the worm, which, like the other cells in the comb, is horizontal. But this habitation remains suitable to the worm called to the royal state only during the first three days of its existence: another situation is requisite for the other two days it is a worm. Then, which is so small a portion of its life, it must inhabit a cell nearly of a pyramidal figure, and hanging perpendicularly; we may say the workers know it; for, after the worm has completed the third day, they prepare the place to be occupied by its new lodging. They gnaw away the cells surrounding the cylindrical tube, mercilessly sacrifice their worms, and use the wax in constructing a new pyramidal tube, which they solder at right angles to the first, and work it downwards. The diameter of this pyramid decreases insensibly from the base, which is very wide, to the point. During the two days that it is inhabited by the worm, a bee constantly keeps its head more or less inserted into the cell, and, when this worker quits it, another comes to occupy its place. In proportion as the worm grows, the bees labour in extending the cell, and bring food, which they place before its mouth, and around its body, forming a kind of cord around it. The worm, which can move only in a spiral direction, turns incessantly to take the food before its head: it insensibly descends, and at length arrives at the orifice of the cell. Now is the time of transformation to a nymph. As any farther care is unnecessary, the bees close the cell with a peculiar substance appropriated for it, and there the worm undergoes both its metamorphoses.

Though M. Schirach supposes that none but worms three days old are selected for the royal treatment, I am certain of the contrary; and that the operation succeeds equally well on those of two days only. I must be permitted to relate at length the evidence I have of the fact, which will both demonstrate the reality of common worms being converted into queens, and the little influence which their age has on the effect of the operation.

I put some pieces of comb, with some workers eggs, in the cells, and of the same kind as those already hatched, into a hive deprived of the queen. The same day several cells were enlarged by the bees, and converted into royal cells, and the worms supplied with a thick bed of jelly. Five were then removed from those cells, and five common worms, which, forty-eight hours before we had seen come from the egg substituted for them. The bees did not seem aware of the change; they watched over the new worms the same as over those chosen by themselves; they continued enlarging the cells, and closed them at the usual time. When they had hatched on them seven days[*], we removed the cells to see the queens that were to be produced. Two were excluded, almost at the same moment, of the largest size, and well formed in every respect. The term of the other cells having elapsed, and no queen appearing, we opened them. In one, was a dead queen, but still a nymph; the other two were empty. The worms had spun their silk coccoons, but died before passing into their nymphine state, and presented only a dry skin. I can

[*] The author's meaning here is obscure.—T.

43

conceive nothing more conclusive than this experiment. It demonstrates that bees have the power of converting the worms of workers into queens; since they succeeded in procuring queens, by operating on the worms which we ourselves had selected. It is equally demonstrated, that the success of the operation does not depend on the worms being three days old, as those entrusted to the bees were only two. Nor is this all; bees can convert worms still younger into queens. The following experiment showed, that when the queen is lost, they destine worms only a few hours old to replace her.

I was in possession of a hive, which being long deprived of the female, had neither egg nor worm. I provided a queen of the greatest fertility; and she immediately began laying in the cells of workers. I removed this female before being quite three days in the hive, and before any of her eggs were hatched. The following morning, that is, the fourth day, we counted fifty minute worms, the oldest scarcely hatched twenty-four hours. However, several were already destined for queens, which was proved by the bees depositing around them a much more abundant provision of food than is supplied to common worms. Next day, the worms were near forty hours old: the bees had enlarged and converted their hexagonal cells into cylindrical ones of the greatest capacity. During the subsequent days, they still laboured at them, and closed them on the fifth from the origin of the worms. Seven days after sealing of the first of these royal cells, a queen of the largest size proceeded from it. She immediately rushed towards the other royal cells, and endeavoured to destroy their nymphs and worms. In another letter, I shall recount the effects of her fury.

From these details, you will observe, Sir, that M. Schirach's experiments had not been sufficiently diversified when he affirmed that it was essential for the conversion of common worms into queens, they should be three days old. It is undoubted, that equal success attends the experiment not only with worms two days old, but also when they have been only a few hours in existence.

After my researches to corroborate M. Schirach's discovery, I was desirous of learning whether, as this observer conceives, the only means which the bees have of procuring a queen, is giving the common worms a certain kind of aliment, and rearing them in the largest cells. You will remember, that M. de Reaumur's sentiments are very different: "The mother should lay, and she does lay, eggs from which flies fit for being mothers must in their turn proceed. She does so; and it is evident the workers know what she is to do. Bees, to which the mother is so precious, seem to take a peculiar interest in the eggs that one is to proceed from, and to consider them of the greatest value. They construct particular cells where they are to be deposited.— The figure of a royal cell only begun, very much resembles a cup, or, more correctly speaking, the cup that has lost its acorn."

M. de Reaumur, though he did not suspect the possibility of a common worm being converted into a queen, conceived that the queen bee laid a particular species of eggs in the royal cells, from which worms should come that would be queens. According to M. Schirach, on the other hand, bees always having the power of procuring a queen by bringing up worms three days old in a particular manner, it would be needless for nature to grant females the faculty of laying royal eggs. Such prodigality is, in his eyes, inconsistent with the ordinary laws of nature. Therefore he maintains, in direct terms, that she does not lay royal eggs in cells purposely prepared to receive them. He considers the royal cells only as common ones, enlarged by the bees at the moment when the included worm is destined for a queen; and adds, that the royal cell would always be too long for the belly of the mother to reach the bottom.

I admit that M. de Reaumur no where says he has seen the queen lay in the royal cell. However he did not doubt the fact;

and, after all my observations, I must esteem his opinion just. It is quite certain that, at particular periods of the year, the bees prepare royal cells; that the females deposit their eggs in them; and that worms, which shall became queens, proceed from these eggs.

M. Schirach's objection, concerning the length of the cells, proves nothing; for the queen does not delay depositing her egg till they are finished. While only sketched and shaped like the cup of an acorn, she lays it. This naturalist, dazzled by the brilliancy of his discovery, saw only part of the truth. He was the first to find out the resource granted to bees by nature, for repairing the loss of their queen; and too soon persuaded himself that she had provided no other resource for the production of females. This error arose from not observing bees in very flat hives: had he used such as mine, he would have found, on opening them in spring, a confirmation of M. de Reaumur's opinion. Then, which is the season of swarming, hives in good condition are governed by a very fruitful queen: there are royal cells of a figure widely different from those constructed around the worms destined by the bees for queens. They are large, attached to the comb by a stalk, and hanging vertically like stalactites, such, in short, as M. de Reaumur has described them. The females lay in them before completion. We have surprised a queen depositing the egg when the cell was only as the cup of an acorn. The workers never lengthen them until the egg has been laid. In proportion as the worm grows, they are enlarged, and closed by the bees when the first transformation approaches. Thus it is true, that, in spring, the queen deposits in royal cells, previously prepared, eggs from which flies of her own species are to come. Nature has, therefore, provided a double means for the multiplication and conservation of their race.

PREGNY, 24. August 1791.

LETTER V.

EXPERIMENTS PROVING THAT THERE ARE SOMETIMES COMMON BEES WHICH LAY FERTILE EGGS.

The singular discovery of M. Riems, concerning the existence of fertile workers, has appeared very doubtful to you, Sir. You have suspected that the eggs ascribed to workers by this naturalist had actually been produced by small queens, which, on account of their size, were confounded with common bees. But you do not positively insist that M. Riems is deceived; and, in the letter which you did me the honour to address to me, you requested me to investigate, by new experiments, whether there are actually working bees capable of laying fertile eggs. I have made these experiments with great care: and it is for you to judge of the confidence they merit.

On the fifth of August 1788, we found the eggs and worms of large drones in two hives, which had both been some time deprived of queens. We also observed the rudiments of some royal cells appended like stalactites to the edges of the combs. The eggs of males were in them. Being perfectly secure that there was no queen of large size among the bees of these two hives, the eggs, which daily became more numerous, were evidently laid either by queens of small size or by fertile workers. I had reason to believe it was actually by common bees, for we had frequently observed them inserting the posterior part into the cells; and assuming the same attitude as the queen when laying. But, not withstanding every exertion, we had never been able to seize one in this situation, to examine it more narrowly. And we were unwilling to assert any thing positively, without having the bees in our hands that had actually laid. Therefore our observations were continued with equal assiduity, in hopes that, by some fortunate chance, or in a moment of address, we could secure one of them. More than a month all our endeavours were abortive.

My assistant then offered to perform an operation that required both courage and patience, and which I could not resolve to suggest, though the same expedient had occurred to myself. He proposed to examine each bee in the hive separately, to discover whether some small queen had not insinuated herself among them, and escaped our first researches. This was an important experiment; for, should no small queen be found, it would be demonstrative evidence that the eggs had been laid by simple workers.

To perform this operation with all possible exactness, immersing the bees was not enough. You know, Sir, that the contact of water stiffens their organs, that it produces a certain alteration of their external figure: and, from the resemblance of small queens to workers, the slightest alteration of shape would prevent us from distinguishing with sufficient accuracy to what species those immersed might belong. Therefore it was necessary to seize the whole bees of both hives, notwithstanding their irritation, and examine their specific character with the utmost care. This my assistant undertook, and executed with great address. Eleven days were employed in it; and, during all that time, he scarcely allowed himself any relaxation, but what the relief of his eyes required. He took every bee in his hand; he attentively examined the trunk, the hind limbs, and the sting: there was not one without the characteristics of the common bee, that is, the little basket on the hind legs, the long trunk, and the straight sting. He had previously prepared glass cases containing combs. Into these, he put each bee after examination. It is superfluous to observe they were confined, which was a precaution indispensible until termination of the experiment. Neither was it enough to establish that the whole were workers; we had also to continue the experiment, and observe whether any would produce eggs. Thus we examined the cells for several days, and soon observed new laid eggs, from which the worms of drones came at the proper time. My assistant held in his hands the bees that produced them; and as he was perfectly certain they were common ones, it is proved that there are sometimes fertile workers in hives.

Having ascertained M. Schirach's discovery, by so decisive an experiment, we replaced all the bees examined, in very thin glass hives, being only eighteen lines thick, and capable of containing but a single row of combs, and thus were extremely favourable to the observer. We thought, by strictly persisting to watch the bees, we might surprise a fertile one in the act of laying, seize and dissect her. This we were desirous of doing, for the purpose of comparing her ovaries with those of queens, and to ascertain the difference. At length, on the eighth of September, we had the good fortune to succeed.

A bee appeared in the position of a female laying. Before she had time to leave the cell, we suddenly opened the hive and seized her. She presented all the external characteristics of common bees; the only difference we could recognise, and that was a very slight one, consisted in the belly seeming less and more slender than that of workers. On dissection, her ovaries were found more fragile, smaller and composed of fewer oviducts than the ovaries of queens. The filaments containing the eggs were extremely fine, and exhibited swellings at equal distances. We counted eleven eggs of sensible size, some of which appeared ripe for laying. This ovary was double like that of queens.

On the ninth of September, we seized another fertile worker the instant she laid, and dissected her. The ovary was still less expanded than that of the preceding bee, and only four eggs had attained maturity. My assistant extracted one from the oviducts, and succeeded in fixing it by an end on a glass slider. We may take this opportunity of remarking, that it is in the oviducts themselves the eggs are imbued with the viscous liquid, with which they are produced, and not in passing through the spherical sac as Swammerdam believed. During the remainder of this month, we found ten fertile workers in the same hives, and dissected them all. In most, the ovaries were easily distinguished, but in some we could not discern the faintest traces of them. In these last, the oviducts to all appearance were but imperfectly developed, and more address than we had acquired in dissection was necessary to distinguish them.

Fertile workers never lay the eggs of common bees; they produce none but those of males. M. Riems had already observed this singular fact; and here all my observations correspond with his. I shall only add to what he says, that fertile workers are not absolutely indifferent in the choice of cells for depositing their eggs. They always prefer large ones; and only use small cells when unable to find those of larger diameter. But they so far correspond with queens whose impregnation has been retarded, that they sometimes lay in royal cells.

Speaking of females laying male eggs alone, I have already expressed my surprise that bees bestow, on those deposited in royal cells, such care and attention as to feed the worms proceeding from them, and, at the period of transformation, to close them up. But I know not, Sir, why I omitted to observe that, after sealing the royal cells, the workers build them up, and sit on them until the last metamorphosis of the included male[*].

---

[*] It is difficult to discover whether the author thinks, as some naturalists, that bees are instrumental in hatching the eggs.—T.

The treatment of the royal cells where fertile workers lay the eggs of drones is very different. They begin indeed with bestowing every care on their eggs and worms; they close the cells at a suitable time, but never fail to destroy them three days afterwards.

Having finished these first experiments with success, I had still to discover the cause of the expansion of the sexual organs of fertile workers. M. Riems had not engaged in this interesting problem; and at first I dreaded that I should have no other guide towards its solution than conjecture. Yet from serious reflection, it appeared, that, by connecting the facts contained in this letter, there was some light that might elucidate my procedure in this new research.

From M. Schirach's elegant discoveries, it is beyond all doubt that common bees are originally of the female sex. They have received from nature the germs of an ovary, but she has allowed its expansion only in the particular case of their receiving a certain aliment while a worm. Thus it must be the peculiar object of inquiry whether the fertile workers get that aliment while worms.

All my experiments convince me that bees, capable of laying, are produced in hives that have lost the queen. A great quantity of royal jelly is then prepared for feeding the larvæ destined to replace her. Therefore, if fertile workers are produced in this situation alone, it is evident their origin is only in those hives where bees prepare the royal jelly. Towards this circumstance, I bent all my attention. It induced me to suspect that when bees give the *royal treatment* to certain worms, they either by accident or a particular instinct, the principle of which is unknown to me, drop some particles of royal jelly into cells contiguous to those containing the worms destined for queens. The larvæ of workers that have accidentally received portions of so active an aliment, must be more or less affected by it; and their ovaries should acquire a degree of expansion. But this expansion will be imperfect; why? because the royal food has been administered only in small portions, and, besides, the larvæ having lived in cells of the smallest dimensions, their parts cannot extend beyond the ordinary proportions. Thus, the bees produced by them will resemble common workers in size and all the external characteristics. Added to that, they will have the faculty of laying some eggs, solely from the effect of the trifling portion of royal jelly mixed with their aliment.

That we may judge of the justness of this explanation, it is necessary to consider fertile workers from their origin; to investigate whether the cells, where they are brought up, are constantly in the vicinity of the royal cells, and if their food is mixed with particles of the royal jelly. Unfortunately, the execution of these experiments is very difficult. When pure, the royal jelly is recognised by its sharp and pungent taste; but, when mixed with other substances, the peculiar savour is very imperfectly distinguished. Thus I conceived, that my investigation should be limited to the situation of the cells; and, as the subject is important, permit me to enter a little into detail[*].

In June 1790, I observed that one of my thinnest hives had wanted the queen several days, and that the bees had no mean of replacing her, there being no workers' worms. I then provided them with a small portion of comb, each cell containing a young worm of the working species. Next day, the bees prolonged several cells around the worms destined for queens, in the form of royal ones. They also bestowed some care on the worms in the adjoining cells. Four days afterwards, all the royal cells were shut, and we counted nineteen small cells also perfected and closed by a covering almost flat. In these were worms that had not received the royal treatment; but as they had lived in the vicinity of the worms destined for replacing the queens, it was

---

[*] The original is extremely confused in the preceding passages.—T.

very interesting to follow their history, and necessary to watch the moment of their last transformation. I removed the nineteen cells into a grated box, which was introduced among the bees. I also removed the royal cells, for it was of great importance, that the queens they would produce should not disturb or derange the result of the experiment. But here another precaution was also requisite. It was to be feared, that the bees being deprived of the produce of their labour, and the object of their hope might be totally discouraged; therefore, I supplied them with another piece of comb, containing the brood of workers, reserving power to destroy the young brood when necessary. This plan succeeded admirably. The bees, in bestowing all their attention on these last worms, forgot those that had been removed.

When the moment of transformation of the nymphs in the nineteen cells arrived, I examined the grated box frequently every day, and at length found six bees exactly similar to *common bees*. The worms of the remaining thirteen had perished without changing.

The portion of brood comb that had been put into the hive to prevent the discouragement of the bees was then removed. I put aside the queens produced in the royal cells; and having painted the thorax of the six bees red, and amputated the right antenna, I transferred the whole six into the hive, where they were well received.

You easily conceive my object, Sir, in this course of observations. I knew there was neither a large nor small queen in the hive: therefore, if, in the sequel, I should find new laid eggs in the combs, how very probable must it be that they had been produced by some of the six bees? But, to attain absolute certainty, it was necessary to take them in the act of laying. Some ineffaceable mark was also required for distinguishing them in particular.

This proceeding was attended with the most ample success. We soon found eggs in the hive; their number increased daily; and their worms were all drones. But a long interval elapsed before we could take the bees that laid them. At length, by means of assiduity and perseverance, we perceived one introducing the posterior part into a cell; we opened the hive, and caught the bee: We saw the egg it had deposited, and by the colour of the thorax, and privation of the right antenna, instantly recognised that it was one of the six that had passed to the vermicular state in the vicinity of the royal cells.

I could no longer doubt the truth of my conjecture; at the same time, I know not whether the truth will appear as rigorous to you, Sir, as it does to myself. But I reason in the following manner: If it is certain that fertile workers are always produced in the vicinity of royal cells, it is no less true, that in itself, the vicinity is indifferent; for the size and figure of these cells can produce no effect on the worms in those surrounding them; there must be something more; we know that a particular aliment is conveyed to the royal cells; we also know, that this aliment has a very powerful effect on the ovaries; that it alone can unfold the germ. Thus, we must necessarily suppose the worms in the adjacent cells have had a portion of the same food. This is what they gain, therefore, by vicinity to the royal cells. The bees, in their course thither, will pass in numbers over them, stop and drop some portion of the jelly destined for the royal larvæ. This reasoning, I presume, is consistent with the principles of sound logic.

I have repeated the experiment now described so often, and weighed all the concomitant circumstances with so much care, that whenever I please, I can produce fertile workers in my hives. The method is simple. I remove the queen from a hive; and very soon the bees labour to replace her, by enlarging several cells, containing the brood of workers, and supplying the included worms with the royal jelly. Portions of this aliment also

54

fall on the young larvæ deposited in the adjacent cells, and it unfolds the ovaries to a certain degree. Fertile workers are constantly produced in hives where the bees labour to replace their queen; but we very rarely find them, because they are attacked and destroyed by the young queens reared in the royal cells. Therefore, to save them, all their enemies must be removed, and the larvæ of the royal cells taken away before undergoing their last metamorphoses. Then the fertile workers, being without rivals at the time of their origin, will be well received, and, by taking the precaution to mark them, it will be seen, in a few days, that they produce the eggs of males. Thus, the whole secret of this proceeding consists in removing the royal cells at the proper time; that is, after being sealed, and previous to the young queens leaving them[*].

I shall add but a few words to this long letter. There is nothing so very surprising in the production of fertile workers, when we consider the consequences of M. Shirach's beautiful discovery. But why do they lay male eggs only? I can conceive, indeed, that the reason of their laying few is from their ovaries being but imperfectly expanded, but I can form no idea why all the eggs should be those of males, neither can I any better account for their use in hives; and hitherto, I have made no experiments on their mode of fecundation.

PREGNY, 25. August 1791.

---

[*] I have frequently seen queens, at the moment of production, begin first by attacking the royal cells and then the common ones beside them. As I had not seen fertile workers when I first observed this fact, I could not conceive from what motive the fury of the queen was thus directed towards the common cells. But now I know they can distinguish the species included, and have the same instinctive jealousy or aversion towards them as against the nymphs of queens properly so denominated.

LETTER VI.

ON THE COMBATS OF QUEENS: THE MASSACRE OF THE MALES: AND WHAT SUCCEEDS IN A HIVE WHERE A STRANGER QUEEN IS SUBSTITUTED FOR THE NATURAL ONE.

M. de Reaumur had not witnessed every thing relative to bees when he composed his history of these industrious animals. Several observers, and those of Lusaçe in particular, have discovered many important facts that escaped him; and I, in my turn, have made various observations of which he had no suspicion: at the same time, and this is a very remarkable circumstance, not only has all that he expressly declares he saw been verified by succeeding naturalists, but all his conjectures are found just. The German naturalists, Schirach, Hattorf, and Riems sometimes contradict him, indeed, in their memoirs; but I can maintain that, while combating the opinion of M. de Reaumur, it is they who are almost always wrong; of which several instances might be adduced.

What I shall now proceed to say will give me an opportunity of detailing some interesting facts.

It was observed by M. de Reaumur, that when any supernumerary queen is either produced in a hive, or comes into it, one of the two soon perishes. He has not actually witnessed the combat in which she falls, but he conjectures there is a mutual attack, and that the empire remains with the strongest or the most fortunate. M. Schirach, on the other hand, and, after him, M. Riems, thinks that the working bees assail the stranger, and sting her to death. I cannot comprehend by what means they have been able to make this observation: as they used very thick hives only, with several rows of combs, they could at most but observe the commencement of hostilities. While the combat lasts, the bees move with great rapidity; they fly on all sides; and, gliding between the combs, conceal their motions from the observer. For my part, though using the most favourable hives, I have never seen a combat between the queens and workers, but I have very often beheld one between the queens themselves.

In one of my hives in particular, there were five or six royal cells, each including a nymph. The eldest first underwent its transformation. Scarcely did ten minutes elapse from the time of this young queen leaving her cradle, when she visited the other royal cells still close. She furiously attacked the nearest; and, by dint of labour, succeeded in opening the top: we saw her tearing the silk of the coccoon with her teeth; but her efforts were probably inadequate to the object, for she abandoned this end of the cell, and began at the other, where she accomplished a larger aperture. When it was sufficiently enlarged, she endeavoured to introduce her belly, and made many exertions until she succeeded in giving her rival a deadly wound with her sting. Then having left the cell, all the bees that had hitherto been spectators of her labour, began to increase the opening, and drew out the dead body of a queen scarcely come from its envelope of a nymph.

Meanwhile, the victorious young queen attacked another royal cell, but did not endeavour to introduce her extremity into it. There was only a royal nymph, and no queen, come to maturity, as in the first cell. In all probability, nymphs of queens inspire their rivals with less animosity; still they do not escape destruction: because, whenever a royal cell has been opened before the proper time, the bees extract the contents in whatever form they may be, whether worm, nymph, or queen. Lastly, the young queen attacked the third cell, but could not succeed in penetrating it. She laboured languidly, and appeared as if exhausted by her first exertions. As we now required queens for some particular experiments, we resolved to remove the other royal cells, yet in safety, to secure them from her fury.

After this observation, we wished to see what ensued on two queens leaving their cells at the same time, and in what manner one perished. I find an observation on this head in my Journal, 15. May 1790.

In one of our thinnest hives, two queens left their cells almost at the same moment. Whenever they observed each other, they rushed together, apparently with great fury, and were in such a position that the antennæ of each was seized by the teeth of the other: the head, breast, and belly of the one were exposed to the head, breast, and belly of the other: the extremity of their bodies were curved; they were reciprocally pierced with the stings; and both fell dead at the same instant. But it seems as if nature has not ordained that both combatants should perish in the duel; but rather that, when finding themselves in the situation described, namely, opposite, and belly to belly, they fly at that moment with the utmost precipitation. Thus, when these two rivals felt the extremities about to meet, they disengaged themselves, and each fled away. You will observe, Sir, that I have repeated this observation very often, so that it leaves no room for doubt: and I think that we may here penetrate the intention of nature.

There ought to be none but one queen in a hive: therefore it is necessary, if by chance a second is either produced or comes into the hive, that one of the two must be destroyed. This cannot be committed to the working bees, because, in a republic composed of so many individuals, an equal consent cannot be supposed always to exist; it might frequently happen that one group of bees destroyed one of the queens, while a second would massacre the other; and the hive thus be deprived of queens. Therefore it was necessary that the queens themselves should be entrusted with the destruction of their rivals: but as, in these combats, nature demands but a single victim, she has wisely arranged that, at the moment when, from their position, the two combatants might lose their lives, both feel so great an alarm, that they think only of flight, and not of using their stings.

I am well aware of the hazard of error in minute researches into the causes of the most trifling facts. But here the object and the means seem so plain, that I have ventured to advance my conjectures. You will judge better than I can, whether they are well founded.—Let me now return from this digression.

A few minutes after the two queens separated, their terror ceased, and they again began to seek each other. Immediately on coming in sight, they rushed together, seized one another, and resumed exactly their former position. The result of this rencounter was the same. When their bellies approached, they hastily disengaged themselves, and fled with precipitation. During all this time, the workers seemed in great agitation; and the tumult appeared to increase when the adversaries separated. Two different times, we observed them stop the flight of the queens, seize their limbs, and retain them prisoners above a minute. At last, the queen, which was either the strongest or the most enraged, darted on her rival at a moment when unperceived, and with her teeth caught the origin of the wing; then rising above her, brought the extremity of her own body under the belly of the other; and, by this means, easily pierced her with the sting. Then she withdrew her sting after losing hold of the wing. The vanquished queen fell down, dragged herself languidly along, and, her strength failing, she soon expired.

This observation proved that virgin queens engage in single combats; but we wished to discover whether those fecundated, and mothers, had the same animosity.

On the 22. of July, we selected a flat hive, containing a very fertile queen: and being curious to learn whether, as virgin queens, she would destroy the royal cells, three were introduced into the middle of the comb. Whenever she observed this, *she* sprung forward on the whole, and pierced them towards the bottom; nor did she desist until the included nymphs were exposed. The workers which had hitherto been spectators of this destruction, now came to carry the nymphs away. They greedily devoured the food remaining at the bottom of the cells, and also sucked the fluid from the abdomen of the nymphs: and then terminated with destroying the cells from which they had been drawn.

In the next place, we introduced a very fertile queen into this hive; after painting the thorax to distinguish her from the reigning queen. A circle of bees quickly formed around the stranger, but their intention was not to caress and receive her well; for they insensibly accumulated so much, and surrounded her so closely, that in scarcely a minute she lost her liberty and became a prisoner. It is a remarkable circumstance, that other workers at the same time collected round the reigning queen and restrained all her motions; we instantly saw her confined like the stranger. Perhaps it may be said, the bees anticipated the combat in which these queens were about to engage, and were impatient to behold the issue of it, for they retained their prisoners only when they appeared to withdraw from each other; and if one less restrained seemed desirous of approaching her rival, all the bees forming the clusters gave way to allow her full liberty for the attack; then if the queens testified a disposition to fly, they returned to enclose them.

We have repeatedly witnessed this fact, but it presents so new and singular a characteristic in the policy of bees, that it must be seen again a thousand times before any positive assertion can be made on the subject. I would therefore recommend that naturalists should attentively examine the combat of queens, and particularly ascertain what part is taken by the workers. Is their object to accelerate the combat? Do they by any secret means excite the fury of the combatants? Whence does it happen that accustomed to bestow every care on their queen, in certain circumstances, they oppose her preparations to avoid impending danger?

A long series of observations are necessary to solve these problems. It is an immense field for experiment, which will afford infinitely curious results. I intreat you to pardon my frequent digressions. The subject is deeply philosophical, genius such as your's is required to treat it properly; and I shall now be satisfied with proceeding in the description of the combat.

The cluster of bees that surrounded the reigning queen having allowed her some freedom, she seemed to advance towards that part of the comb where her rival stood; then, all the bees receded before her, the multitude of workers, separating the two adversaries, gradually dispersed, until only two remained; these also removed, and allowed the queens to come in sight. At this moment, the reigning queen rushed on the stranger, with her teeth seized her near the origin of the wing, and succeeded in fixing her against the comb without any possibility of motion or resistance. Next curving her body, she pierced this unhappy victim of our curiosity with a mortal wound.

In the last place, to exhaust every combination, we had still to examine whether a combat would ensue between two queens, one impregnated, and the other a virgin; and what circumstances attended it.

On the 18. of September, we introduced a very fruitful queen into a glass hive, already containing a virgin queen, and put her on the opposite side of the comb, that we might have time to see how the workers would receive her. She was immediately surrounded, but they confined her only a moment. Being oppressed with the necessity of laying, she dropped some eggs; however, we could not discover what became of them; certainly the bees did not convey them to the cells, for, on inspection, we found none there. The group surrounding this queen having dispersed a little, she advanced towards the edge of the comb, and soon approached very near the virgin queen. When in sight, they rushed together; the virgin queen got on the back of the other, and gave her several stings in the belly, but, having aimed at the scaly part, they did not injure her, and the combatants separated. In a few minutes, they returned to the charge; but this time the impregnated queen mounted on her rival; however, she sought in vain to pierce her, for the sting did not enter; the virgin queen then disengaged herself and fled; she also succeeded in escaping another attack, where her adversary had the advantage of position. These rivals appeared nearly of equal strength; and it was difficult to foresee to which side victory would incline, until at last, by a successful exertion, the virgin queen mortally wounded the stranger, and she expired in a moment. The sting had penetrated so far that the victorious queen was unable to extract it, and she was overthrown by the fall of her enemy. She made great exertions to disengage the sting: but could succeed by no other means than turning on the extremity of the belly, as on a pivot. Probably the barbs of the sting fell by this motion, and, closing like a spiral around the stem, came more easily from the wound.

These observations, Sir, I think will satisfy you, respecting the conjecture of our celebrated Reaumur. It is certain, that if several queens are introduced into a hive, one alone will preserve the empire; that the others will perish from her attacks; and that the workers will at no time attempt to employ their stings against the stranger queen. I can conceive what has misled M. Riems and Schirach; but it is necessary for explaining it that I should relate a new feature in the policy of bees, at considerable length.

In the natural state of hives, several queens from different royal cells, may sometimes exist at the same moment, and they will remain either until formation of a swarm or a combat among them decides to which the throne shall appertain. But excepting this case, there never can be supernumerary queens; and if an observer wishes to introduce one, he can accomplish it only by force, that is by opening the hive. In a word, no queen can insinuate herself into a hive in a natural state, for the following reasons.

Bees preserve a sufficient guard, day and night, at the entrance of their habitation. These vigilant centinels examine whatever is presented; and, as if distrusting their eyes, they touch with the antennæ every individual endeavouring to penetrate the hive, and also the various substances put within their reach; which affords us an opportunity of observing that the antennæ are certainly the organs of feeling. If a stranger queen appears, she is instantly seized by the bees on guard, which prevent her entry by laying hold of her legs or wings with their teeth, and crowd so closely around her, that she cannot move. Other bees, from the interior of the hive, gradually come to their assistance, and confine her still more narrowly, all having their heads towards the centre where the queen is inclosed; and they remain with such evident anxiety, eagerness, and attention, that the cluster they form may be carried about for some time, without their being sensible of it. A stranger queen, so closely confined and hemmed in, cannot possibly penetrate the hive. If the bees retain her too long imprisoned, she perishes. Her death probably ensues from hunger, or the privation of air; it is undoubted, at least, that she is never stung. We never saw the bees direct their stings against her, except a single time, and then it was owing to ourselves. We endeavoured, from compassion for a queen's situation, to remove her from the centre of a cluster; the bees became enraged; and, in darting out their stings, some struck the queen, and killed her. It is so certain that the stings were not purposely directed against her, that several of the workers were themselves killed; and surely they could not intend destroying one another. Had we not interfered, they would have been content with confining the queen, and would not have massacred her.

It was in similar circumstances that M. Riems saw the workers anxiously pursue a queen. He thought they designed to sting her, and thence concluded, that the office of the common bees is to kill supernumerary queens. You have quoted his observations in the *Contemplation de la Nature, part II, chap. 27, note 7*. But you are sensible, Sir, from these details, that he has been mistaken. He did not know the attention that bees bestow on what passes at the entrance of their hive, and he was entirely ignorant of the means they take to prevent supernumerary queens from penetrating it.

◆　　◆　　◆　　◆　　◆

After ascertaining that the workers in no situation sting the supernumerary queens, we were curious to learn how a stranger queen would be received in a hive wanting a reigning one. To elucidate this matter, we made numerous experiments, the detail of which would protract this letter too much, therefore I shall relate only the principal results.

Bees do not immediately observe the removal of their queen; their labours are uninterrupted; they watch over the young, and perform all their ordinary occupations. But, in a few hours, agitation ensues; all appears a scene of tumult in the hive. A singular humming is heard; the bees desert their young; and rush over the surface of the combs with a delirious impetuosity. Then they discover their queen is no longer among them. But how do they become sensible of it? How do the bees on the surface of the comb discover that the queen is not on the next comb? In treating of another characteristic of these animals, you have yourself, Sir, proposed the same question; I am incapable of answering it indeed, but I have collected some facts, that may perhaps facilitate the elucidation of this mystery.

I cannot doubt that the agitation arises from the workers having lost their queen; for on restoring her, tranquillity is instantly regained among them; and, what is very singular, they *recognise* her: you must interpret this expression strictly. Substitution of another queen is not attended with the same effect, if she is introduced into the hive within the first twelve hours after removal of the reigning one. Here the agitation continues; and the bees treat the stranger the same as when the presence of their own leaves them nothing to desire. They surround, seize, and keep her captive, a very long time, in an impenetrable cluster; and she commonly dies either from hunger or privation of air.

If eighteen hours elapse before substitution of a stranger queen for the native one removed, she is at first treated in the same manner, but the bees leave her sooner; nor is the surrounding cluster so close; they gradually disperse; and the queen is at last liberated. She moves languidly; and sometimes expires in a few minutes. However some queens have escaped in good health from an imprisonment of seventeen hours; and ended with reigning in the hives where they had originally been ill received.

If, before substituting the stranger queen, twenty-four hours elapse, she will be well received, and reign from the moment of her introduction into the hive. Here I speak of the good reception given to a queen after an interregnum of twenty-four hours. But as this word reception is very indefinite, it is proper to enter into some detail for explaining the exact sense in which I use it. On the 15. of August, I introduced a fertile queen, eleven months old, into a glass hive. The bees were twenty-four hours deprived of their queen, and had already begun the construction of twelve royal-cells, such as described in the preceding chapter. Immediately on placing this female stranger on the comb, the workers near her touched her with their antennæ, and, passing their trunks over every part of her body, they gave her honey. Then these gave place to others that treated her exactly in the same manner. All vibrated their wings at once, and ranged themselves in a circle around their sovereign. Hence resulted a kind of agitation which gradually communicated to the workers situated on the same surface of the comb, and induced them to come and reconnoitre, in their turn, what was going on. They soon arrived; and, having broke through the circle formed by the first, approached the queen, touched her with the antennæ, and gave her honey. After this little ceremony they retired; and, placing themselves behind the others, enlarged the circle. There they vibrated their wings, and buzzed without tumult or disorder, and as if experiencing some very agreeable sensation. The queen had not yet moved from the place where I had put her, but in a quarter of an hour she began to move. The bees, far from opposing her, opened the circle at that part to which she turned, followed her, and formed a guard around. She was oppressed with the necessity of laying, and dropped eggs. Finally, after four hours abode, she began to deposit male eggs in the cells she met.

While these events passed on the surface of the comb where the queen stood, all was quiet on the other side. Here the workers were apparently ignorant of a queen's arrival in the hive. They laboured with great activity at the royal cells, as if ignorant that they no longer stood in need of them: they watched over the royal worms, supplied them with jelly and the like. But the queen having at length come to this side, she was received with the same respect that she had experienced from their companions on the other side of the comb. They encompassed her; gave her honey; and touched her with their antennæ: and what proved better that they treated her as a mother, was their immediately desisting from work at the royal cells; they removed the worms, and devoured the food collected around them. From this moment the queen was recognised by all her people, and conducted herself in this new habitation as if it had been her native hive.

These particulars will give a just idea of the manner that bees receive a stranger queen; when they have time to forget their own, she is treated exactly as if she was their natural one, except that there is perhaps at first greater interest testified in her, or more conspicuous demonstrations of it. I am sensible of the impropriety of these expressions, but M. de Reaumur in some respect authorises them. He does not scruple to say, that bees pay *attention*, *homage*, and *respect*, to their queen, and from his example the like expressions have escaped most authors that treat on bees.

Twenty-four or thirty hours absence is sufficient to make them forget their first queen, but I can hazard no conjecture on the cause.

◆   ◆   ◆   ◆   ◆

Before terminating this letter, which is full of combats and disastrous scenes, I should, perhaps, give you an account of some more pleasing and interesting facts relative to their industry. However, to avoid returning to duels and massacres, I shall here subjoin my observations on the massacre of the males.

You will remember, Sir, it is agreed by all observers, that at a certain period of the year, the workers kill and expel the drones. M. de Reaumur speaks of these executions as a horrible massacre. He does not expressly affirm, indeed, that he has himself witnessed it, but what we have seen corresponds so well with his account, that there can be no doubt he has beheld the peculiarities of the massacre.

It is usually in the months of July and August, that the bees free themselves of the males. Then they are drove away and pursued to the inmost parts of the hive, where they collect in numbers; and as at the same time we find many dead drones on the ground before the hives, it is indubitable that after being expelled, the bees sting them to death. Yet on the surface of the comb, we do not see the sting used against them; there the bees are content to pursue and drive them away. You observe this, Sir, yourself, in the new notes added to *la Contemplation de la Nature*; and you seem disposed to think, that the drones forced to retire to the extremity of the hive, perish from hunger. Your conjecture was extremely probable. Still it was possible the carnage might take place in the bottom of the hive, and had been unobserved, because that part is dark, and escapes the observer's eye.

To appreciate the justice of this suspicion, we thought of making the support of the hive of glass, and of placing ourselves below to see what passed in the scene of action. Therefore, a glass table was constructed, on which were put six hives with swarms of the same year; and, lying under it, we endeavoured to discover how the drones were destroyed. The invention succeeded to admiration. On the 4 of July, we saw the workers actually massacre the males, in the whole six swarms, at the same hour, and with the same peculiarities.

The glass table was covered with bees full of animation, which flew upon the drones, as they came from the bottom of the hive; seized them by the antennæ, the limbs, and the wings, and after having dragged them about, or, so to speak, after quartering them, they killed them by repeated stings directed between the rings of the belly. The moment that this formidable weapon reached them, was the last of their existence; they stretched their wings, and expired. At the same time, as if the workers did not consider them as dead as they appeared to us, they still stuck the sting so deep, that it could hardly be withdrawn, and these bees were obliged to turn upon themselves before the stings could be disengaged.

Next day, having resumed our former position, we witnessed new scenes of carnage. During three hours, the bees furiously destroyed the males. They had massacred all their own on the preceding evening, but now attacked those which, driven from the neighbouring hives, had taken refuge amongst them. We saw them also tear some remaining nymphs from the combs; they greedily sucked all the fluid from the abdomen, and then carried them away. The following days no drones remained in the hives.

These two observations seem to me decisive. It is incontestible that nature has charged the workers with the destruction of the males at certain seasons of the year. But what means does she use to excite their fury against them? This is a question that I cannot pretend to answer. However, an observation I have made may one day lead to solution of the problem. The males are never destroyed in hives deprived of queens, on the contrary, while a savage massacre prevails in other places, they there find an asylum. They are tolerated and fed, and many are seen even in the middle of January. They are also preserved in hives, which, without a queen properly so called, have some individuals of that species that lay the eggs of males, and in those whose half fecundated queens, if I may use the expression, propagate only drones. Therefore, the massacre takes place but in hives where the queens are completely fertile, and it never begins until the season of swarming is past.

PREGNY, 28 August 1791.

LETTER VII.

SEQUEL OF EXPERIMENTS ON THE RECEPTION OF A STRANGER QUEEN. M. DE REAUMUR'S OBSERVATIONS ON THE SUBJECT.

I have frequently testified my admiration of M. de Reaumur's observations on bees. I feel a sensible pleasure in acknowledging that if I have made any progress in the art of observation, I am indebted for it to profound study of the works of this naturalist. In general his authority has such weight, that I can scarcely trust my own experiments when the results are different from his. Likewise, on finding myself in opposition to the *historian of bees*, I repeat my experiments. I vary the mode of conducting them; I examine with the utmost caution all the circumstances that might mislead me, and never are my labours interrupted before acquiring the moral certainty of avoiding error. With the aid of these precautions, I have discovered the justice of M. de Reaumur's suggestions, and I have a thousand times seen, if certain experiments seemed to combat them, it was from incorrectness of execution. Yet I must except some facts where my results have constantly been different from his. Those respecting the reception of a stranger queen substituted for the natural one, are of the number.

If, after removing the natural queen, a stranger is immediately substituted, the usurper is ill received. I never could succeed in making them adopt her, but by allowing an interval of twenty or twenty-four hours to elapse. Then they seemed to have forgot their own queen; and respectfully received any female put in her place. M. de Reaumur, on the contrary, asserts, that should the original queen be removed, and another presented, this new one will be perfectly well received from the beginning. As evidence of this assertion, he gives the detail of an experiment which must be read in his work, for I shall here give only an extract of it[*]. He induced four or five hundred bees to leave their native hive and enter a glass box, containing a small piece of comb towards the top. At first they were in great agitation; and, to pacify or console them, he presented a new queen. From this moment, the tumult ceased, and the stranger queen was received with all respect.

I do not dispute the truth of this experiment; but, in my opinion, it does not warrant the conclusion that M. de Reaumur deduces from it. His apparatus removed the bees too much from their natural condition, to allow him to judge of their instinct and dispositions. In other situations, he has himself observed, that these animals, reduced to small numbers, lost their industry and activity, and feebly continued their ordinary labours. Thus their instinct is affected by every operation that too much diminishes their number. To render such an experiment truly conclusive, it must be made in a populous hive; and on removing the native queen, a stranger must immediately be substituted in her place. Had this been done, I am fully persuaded, that M. de Reaumur would have seen the bees imprison the usurper, confine her at least twelve or fifteen hours among them, and frequently suffocate her: nor would he have witnessed any favourable reception before an interval of twenty-four hours after removal of the original queen. No variation has occurred in my experiments regarding this fact. Their number, and the attention bestowed on them, make me presume they merit your confidence.

---

[*] Edit. 4to, Tom. V. p. 258.

M. de Reaumur, in another passage of the same Memoir, affirms, that *bees, which have a queen they are satisfied with, are nevertheless disposed to give the best possible reception to any female that seeks refuge among them.* In the preceding letter, I have related my experiments on this head: their success has been very different from that of M. de Reaumur's. I have proved that the workers never employ their stings against the queen; but this cannot be called the welcome reception of a stranger. They retain her within their ranks, and seem to allow her liberty only when she prepares to combat the reigning queen. This observation cannot be made except in the thinnest hives. Those used by M. de Reaumur had always two parallel combs at least, which must have prevented him from observing some very important circumstances that influence the conduct of workers when supplied with several females. The first circles formed around a stranger queen he has taken for caresses; and, from the little that this queen could advance between the combs, it must have been impossible for him to observe that the circles, which always continued contracting, ended in restraint of the females there inclosed. Had he used thinner hives, he would have discovered that what he supposed indication of a favourable reception was the prelude of actual imprisonment.

I feel reluctant to assert that M. de Reaumur was deceived. Yet I cannot admit that, on certain occasions, bees tolerate a plurality of females in their hives. The experiment on which this affirmation rests will not be considered decisive. In the month of December, he introduced a stranger queen into a glass hive, in his cabinet, and confined her there. The bees had no opportunity of going out. This stranger was well received; her presence awakened the workers from their lethargic state, into which they did not relapse; she excited no carnage; the number of dead bees on the board of the hive did not sensibly increase; and no dead queens were found.

Before concluding any thing favourable to the plurality of queens, it was necessary to ascertain whether the native queen was living when the new one was introduced into the hive: however the author neglected this; and it is very probable the hive had lost its queen, since the bees were languid, and the presence of a stranger restored their activity.

I trust, Sir, that you will pardon this slight criticism. Far from industriously seeking faults in our celebrated Reaumur, I derive the greatest pleasure when my observations coincide with his, and still more, when my experiments justify his conjectures. But I think it proper to point out those cases where the imperfections of his hives have led him into error, and to explain from what causes I have not seen certain facts in the same manner he did. I feel particular anxiety to merit your confidence, and I am aware that the greatest exertions are necessary, when I have to combat the historian of bees. I confide in your judgment; and pray you to be assured of my respect.

PREGNY, 30. August 1791.

LETTER VIII.

IS THE QUEEN OVIPAROUS? WHAT INFLUENCE HAS THE SIZE OF THE CELLS, WHERE THE EGGS ARE DEPOSITED, ON THE BEES PRODUCED?—RESEARCHES ON THE MODE OF SPINNING THE COCCOONS.

In this letter I shall collect some isolated observations relative to various points in the history of bees, concerning which you wished me to engage.

You desired me to investigate whether the queen is really *oviparous*. M. de Reaumur leaves this question undecided. He observes, that he has never seen the worm hatched; and he only asserts that worms are found in those cells where eggs have been deposited three days preceding. If we attempt to catch the moment when the worm leaves the egg, we must extend our observations beyond the interior of the hive; for there the continual motion of the bees obscures what passes at the bottom of cells. The egg must be taken out, presented to the microscope, and every change attentively watched. One other precaution is essential. As a certain degree of heat is requisite to hatch the worms, should the eggs be too soon deprived of it they wither and perish. The sole method of succeeding in seeing the worm come out, consists in watching the queen while she lays, in marking the egg so as to be recognised, and removing it from the hive to the microscope only an hour or two before the three days elapse. The worm will certainly be hatched, provided it has been exposed as long as possible to the full degree of heat. Such is the course I have pursued; and the following are the results obtained.

In the month of August, we removed several cells containing eggs that had been three days deposited: we cut off the top of the cell, and put the pyramidal bottom, where the egg was fixed, on a glass slider. Slight motions were soon perceptible in the eggs. At first, we could observe no external organization: the worm was entirely concealed from us by its pellicle. We then prepared to examine the egg with a powerful magnifier; however, during the interval, the worm burst its surrounding membrane, and cast off part of the envelope, which was torn and ragged on different parts of the body, and more evidently so towards the last rings. The worm alternately curved and stretched itself, with very lively action. Twenty minutes were occupied in casting off the spoil; when this exertion ceased: the worm lay down, curved, and seemed to take that rest which it required. An egg laid in a worker's cell produced this animal, which would have become a worker itself.

We next directed our attention to the moment when a male worm would be hatched. An egg was exposed to the sun on a glass slider; and, with a good magnifier, nine rings of the worm were perceptible within the transparent pellicle. This membrane was still entire, and the worm perfectly motionless. The two longitudinal lines of tracheæ were visible on the surface, and many ramifications. We never lost sight of the egg a single instant, and now succeeded in observing the first motions of the worm. The thick end alternately straightened and curved, and almost reached the part where the sharp extremity was fixed. These exertions burst the membrane, first on the upper part, towards the head, then on the back, and afterwards on all the rest successively. The ragged pellicle remained in folds on different parts of the body, and then fell off. Thus it is beyond dispute, that the queen is oviparous.

Some observers affirm, that the workers attend to the eggs before the worms are hatched; and it is certain that, at whatever time a hive is examined, we always see some workers with the head and thorax inserted into cells containing eggs, and remaining motionless several minutes in this position. It is impossible to discover what they do, for the interior of the cell is concealed by their bodies; but it is very easily ascertained that, in this attitude, they are doing nothing to the eggs.

If, at the moment the queen lays, her eggs are put into a grated box, and deposited in a strange hive, where there is the necessary degree of heat, the worms come out at the usual time, just as if they had been left in the cells. Thus no extraordinary aid or attention is required for their exclusion.

When the workers penetrate the cells, and remain fifteen or twenty minutes motionless, I have reason to believe, it is only to repose from their labours. My observations on the subject seem correct. You know, Sir, that a kind of irregular shaped cells, are frequently constructed on the panes of the hive. These, being glass on one side, are exceedingly convenient to the observer, since all that passes within is exposed. I have often seen bees enter these cells when nothing could attract them. The cells contained neither eggs nor honey, nor did they need further completion. Therefore the workers repaired thither only to enjoy some moments of repose. Indeed, they were fifteen or twenty minutes so perfectly motionless, that had not the dilatation of the rings shewed their respiration, we might have concluded them dead. The queen also sometimes penetrates the large cells of the males, and continues very long motionless in them. Her position prevents the bees from paying their full homage to her, yet even then the workers do not fail to form a circle around her, and brush the part of her belly that remains exposed.

The drones do not enter the cells while reposing, but cluster together on the combs; and sometimes retain this position eighteen or twenty hours without the slightest motion.

As it is important, in many experiments, to know the exact time that the three species of bees exist before assuming their ultimate form, I shall here subjoin my own observations on the point.

The worm of workers passes three days in the egg, five in the vermicular state, and then the bees close up its cell with a wax covering. The worm now begins spinning its coccoon, in which operation thirty-six hours are consumed. In three days, it changes to a nymph, and passes six days in this form. It is only on the twentieth day of its existence, counting from the moment the egg is laid, that it attains the fly state.

The royal worm also passes three days in the egg, and is five a worm; the bees then close its cell; and it immediately begins spinning the coccoon, which occupies twenty-four hours. The tenth and eleventh day it remains in complete repose, and even sixteen hours of the twelfth. Then the transformation to a nymph takes place, in which state four days and a third are passed. Thus it is not before the sixteenth day that the perfect state of queen is attained.

The male worm passes three days in the egg, six and a half as a worm, and metamorphoses into a fly on the twenty-fourth day after the egg is laid.

Though the larvæ of bees are apodal, they are not condemned to absolute immobility in their cells; for they can move by a spiral motion. During the first three days, this motion is so slow as scarcely to be perceptible, but it afterwards becomes more evident. I have then observed them perform two complete revolutions in an hour and three quarters. When the period of transformation arrives, they are only two lines from the orifice of the cells. As their position is constantly the same, bent in an arc, those in the workers' and drones' cells are perpendicular to the horizon, while those in the royal cells lie horizontally. It might be thought, that the difference of position has much influence on the increment of the different larvæ; yet it has none. By reversing combs containing common cells full of brood, I have put the worms in a horizontal position; but they were not injured. I have also turned the royal cells, so that the worms came into a horizontal direction; however their increment was neither slower nor less perfect.

◆   ◆   ◆   ◆   ◆

I have been much surprised at the mode of bees spinning their coccoons, and I have witnessed many new and interesting facts. The worms both of workers and males fabricate *complete* coccoons in their cells; that is, close at both ends, and surrounding the whole body. The royal larvæ, on the other hand, spin imperfect coccoons, open behind, and enveloping only the head, thorax, and first ring of the abdomen. The discovery of this difference, which at first may seem trifling, has given me extreme pleasure, for it evidently demonstrates the admirable art with which nature connects the various characteristics in the industry of bees.

You will remember, Sir, the evidence I gave you of the mutual aversion of queens, of the combats in which they engage, and the animosity that leads them to destroy one another. Of several royal nymphs in a hive, the first transformed attacks the rest, and stings them to death. But were these nymphs enveloped in a complete coccoon, she could not accomplish it. Why? because the silk is of so close a texture, the sting could not penetrate, or if it did, the barbs would be retained by the meshes of the coccoon, and the queen, unable to retract it, would become the victim of her own fury. Thus, that the queen might destroy her rivals, it was necessary the last rings of the body should remain uncovered, therefore the royal nymphs must only form imperfect coccoons. You will observe, that the last rings alone should be exposed, for the sting can penetrate no other part: the head and thorax are protected by connected shelly plates which it cannot pierce.

Hitherto, philosophers have claimed our admiration of nature in her care of preserving and multiplying the species. But from the facts I relate, we must admire her precautions in exposing certain individuals to a mortal danger.

The detail on which I have just entered clearly indicates the final cause of the opening left by the royal worms in their coccoons; but it does not shew whether it is in consequence of a particular instinct that they leave this opening, or whether the wideness of their cells prevents them from stretching the thread up to the top. This question interested me very much; the only method of deciding it was to observe the worms while spinning, which cannot be done in their opaque cells. It then occurred to me to dislodge them from their own habitations, and introduce them into glass tubes, blown in exact imitation of the different kind of cells. The most difficult part of the operation consisted in extracting worms and introducing them here; but my assistant accomplished it with much address. He opened several sealed royal cells, where we knew the larvæ were about to begin their coccoons, and, taking them gently out, introduced one into each of my glass cells without the smallest injury.

They soon prepared to work; and commenced by stretching the anterior part of the body in a straight line, while the other was bent in a curve. This formed a curve of which the longitudinal sides of the cells were tangents, and afforded two points of support. The head was next conducted to the different parts of the cell which it could reach, and it carpeted the surface with a thick bed of silk. We remarked that the threads were not carried from one side to another, and that this would have been impracticable, for the worms being obliged to support themselves, and to keep the posterior rings curved, the free and moveable part of the body was not long enough for the mouth to reach the sides diametrically opposite, and fix the threads to them. You will remember, Sir, that the royal cells are of a pyramidal form, with a wide base, and a long contracted top. These cells hang perpendicularly in the hive, the point downwards, from which position the royal worm can be supported in the cell, only when the curvature of the posterior part forms two points of support; and that it cannot obtain this support without resting on the lower part, or towards the extremity. Therefore if it attempted to stretch out and spin towards the wide end of the cell, it could not reach both sides from being too distant. One part would be touched by its extremity, the other by its back, and it would consequently tumble down. I have particularly ascertained the fact in glass cells that were too large, and of which the diameter was greater towards the point than is usual in cells; there they were unable to support themselves.

These first experiments obviated the suspicion of any particular instinct in the royal worms. They proved, if the worms spun incomplete coccoons, it was because they were forced to do so by the figure of their cells. However, I wished to have evidence still more direct. I put them into cylindrical glass cells, or portions of glass tubes resembling common cells, and I had the satisfaction of seeing them spin complete coccoons, as the worms of workers do. Lastly, I put common worms in very wide cells, and they left the coccoon open. Thus it is demonstrated, that the royal worms, and those of workers, have the same instinct and the same industry, or in other words, when situated in the same circumstances, the course they follow is the same. I may here add, that the royal worms artificially lodged in cells, where they can spin complete coccoons, undergo all their metamorphoses equally well. Thus the necessity imposed on them by nature, of having the coccoons open, is not necessary for their increment; nor has it any other object than that of exposing them to the certainty of perishing by the wounds of their natural enemy; an observation new and truly singular.

◆　　◆　　◆　　◆　　◆

I ought to relate my experiments on the influence that the size of the cells has on bees. It is to you, Sir, that I am indebted for suggesting them.

As we sometimes find males smaller than they ought to be, and also queens more diminutive than usual, it was desirable to obtain a general explanation, to what degree the cells, where bees pass the first period of their existence, influence their size. With this view, you have advised me to remove all the combs composed of common cells, and to leave those consisting of large cells only. It was evident if the common eggs which the queen would lay in these large cells produced workers of larger size, we were bound to conclude that the size of the cells had a sensible influence on the size of the bees. The first time I made this experiment, it did not succeed, because weevils lodged in the hive discouraged the bees. But I repeated it afterwards, and the result was very remarkable.

I removed the whole comb, consisting of common cells, from one of my best glass hives, and left that composed of males' cells alone: and to avoid vacuities, I supplied others of the same kind. This was in June, the season most favourable to bees. I expected that the bees would quickly have repaired the ravages produced by this operation in their dwelling; that they would labour at the breaches, and unite the new combs to the old. But I was very much surprised to see that they did not begin to work. Expecting they would resume their activity, I continued observing them several days; however, my hopes were disappointed. Their homage to the queen was not interrupted indeed; but except in this, their conduct to the queen was quite different from what it usually is; they clustered on the combs without exciting any sensible heat. A thermometer among them rose only to 81°, though standing at 77° in the open air. In a word, they appeared in a state of the greatest despondency.

The queen herself, though very fertile, and though she must have been oppressed by her eggs, hesitated long before depositing them in the large cells; she chose rather to drop them at random than lay in cells unsuitable. However, on the second day, we found six that had been deposited there with all regularity. The worms were hatched three days afterwards, and then we began to study their history. Though the bees provided them with food, they did not carefully attend to it; yet I was in hopes they might be reared. I was again disappointed; for next morning all the worms had disappeared, and their cells were left empty. Profound silence reigned in the hive; few bees left it, and these returned without pellets of wax on the limbs; all was cold and inanimate. To promote a little motion, I thought of supplying the hive with a comb, composed of large cells, full of male brood of all ages. The bees, which had twelve days obstinately refused working in wax, did not unite this comb to their own. However, their industry was awakened in a way that I had not anticipated. They removed all the brood from this comb, cleaned out the whole cells, and prepared them for receiving new eggs. I cannot determine whether they expected the queen to lay, but it is certain if they did so they were not deceived. From this moment, she no longer dropped her eggs; but laid such a number in the new comb, that we found five or six together in the same cell. I then removed all the combs composed of large cells to substitute small cells in their place, an operation which restored complete activity among the bees.

The peculiarities of this experiment seem worthy of attention. It proves that nature does not allow the queen the choice of the eggs she is to lay. It is ordained that, at a certain time of the year, she shall produce those of males, and at another time the eggs of workers, and this order cannot be inverted. We have seen that another fact led me to the same consequence; and as that was extremely important, I am delighted to have it confirmed by a new observation. Let me repeat, therefore, that the eggs are not indiscriminately mixed in the ovaries of the queen, but arranged so that, at a particular season, she can lay only a certain kind. Thus, it would be vain at that time of the year, when the queen should lay the eggs of workers, to attempt forcing her to lay male eggs, by filling the hives with large cells; for, by the experiment just described, we learn, that she will rather drop the workers eggs by chance than deposit them in an unsuitable place; and that she will not lay the eggs of males. I cannot yield to the pleasure of allowing this queen discernment or foresight, for I observe a kind of inconsistency in her conduct. If she refused to lay the eggs of workers in large cells, because nature has instructed her that their size is neither proportioned to the size nor necessities of common worms, would not she also have been instructed not to lay several eggs in one cell? It seems much easier to rear a worker's worm in a large cell, than to rear several of the same species in a small one. Therefore, the supposed discrimination of bees is not very conspicuous. Here the most prominent feature of industry appears in the common bees. When I supplied them with a comb of small cells, full of male brood, their activity was awakened; but instead of bestowing the necessary care on this brood, as they would have done in every other situation, they destroyed the whole nymphs and larvæ, and cleaned out their cells, that the queen, now oppressed with the necessity of laying, might suffer no delay in depositing her eggs. Could we allow them either reason or reflection, this would be an interesting proof of their affection for her.

The experiment, now detailed at length, not having fulfilled my object in determining the influence of the size of the cells on that of the worms, I invented another which proved more successful.

Having selected a comb of large cells, containing the eggs and worms of males, I removed all the worms from their farina, and my assistant substituted those of workers a day old in their place. Then he introduced this comb into a hive that had the queen. The bees did not abandon these substituted worms; they covered their cells with a top almost flat, a kind quite different from what is put on the cells of males; which proves, that they were well aware that these, though inhabiting large cells, were not males. This comb remained eight days in the hive, counting from the time the cells were sealed. I then removed it to examine the included nymphs, which proved those of workers in different stages of advancement; but, as to size and figure, they perfectly resembled what had grown in the smallest cells. I thence concluded, that the larvæ of workers do not acquire greater size in large than in small cells. Although this experiment was made only once, it seems decisive. Nature has appropriated cells of certain dimensions for the worms of workers while in their vermicular state; undoubtedly she has ordained that their organs should be fully expanded, and there is sufficient space for that purpose; therefore more would be useless. Their expansion ought to be no greater in the most spacious cells than in those appropriated for them. If some cells smaller than common ones are found in combs, and the eggs of workers are deposited there, the size of the bees will probably be less than that of common workers, because they have been cramped in the cells; but it does not thence ensue, that a larger cell will admit of them growing to a greater size.

The effect produced on the size of drones by the size of the cells their worms inhabit, may serve as a rule for what should happen to the larvæ of workers in the same circumstances. The large cells of males are sufficiently capacious for the perfect expansion of their organs. Thus, although reared in cells of still greater capacity, they will grow no larger than common drones. We have had evidence of this in those produced by queens whose fecundation has been retarded. You will remember, Sir, that they sometimes lay male eggs in the royal cells. Now, the males proceeding from them, and reared in cells much more spacious than nature has appropriated for them, are no larger than common males. Therefore it is certain, that whatever be the size of the cells where the worms acquire their increment, the bees will attain no greater size than is peculiar to their species. But if, in their primary form, they live in cells smaller than they should be, as their growth will be checked, they will not attain the usual size, of which there is proof in the following experiment. I had a comb consisting of the cell of large drones, and one with those of workers, which also served for the male worms. Of these, my assistant took a certain number from the smallest cells, and deposited them on a quantity of food purposely prepared in the large ones; and in return he introduced into the small cells the worms that had been hatched in the other, and then committed both to the care of the workers in a hive where the queen laid the eggs of males only. The bees were not affected by this change; they took equal care of the worms; and when the period of metamorphosis arrived, gave both kinds that convex covering usually put on those of the males. Eight days afterwards, we removed the combs, and found, as I had expected, nymphs of large males in the large cells, and those of small males in the small ones.

You suggested another experiment which I carefully made, but it met with an unforeseen obstacle. To appreciate the influence of the royal food on the expansion of the worms, you desired me to supply the worm of a worker in a common cell with it. Twice I have attempted this operation without success. Nor do I think it can ever succeed. If bees get the charge of worms, in whose cells the royal food is deposited, and if at the same time they have a queen, they soon remove the worms and greedily devour the food. When, on the contrary, they are deprived of a queen, they change the cells containing worms into cells of the largest kind. Then the worms will infallibly be converted to queens.

But there is another situation where we can judge of the influence of the royal food administered to worms in common cells. I have spoken at great length in my letter on the existence of fertile workers. You cannot forget, Sir, that the expansion of their sexual organs is owing to the reception of some particles of royal jelly, while in the vermicular form. For want of new observations, I must refer you to what is previously said on the subject.

PREGNY, 4 September 1791.

LETTER IX.

ON THE FORMATION OF SWARMS.

I can add but a few facts to the information M. de Reaumur has communicated relative to swarms.

A young queen, according to this celebrated naturalist, is always or almost always at the head of a swarm; but he does not assert the fact positively, and had some doubts on the subject. "Is it certain," says he "as we have hitherto supposed, in coincidence with all who have treated of bees, that the new colony is always conducted by a young mother? May not the old mother be disgusted with her habitation? or may she not be influenced by some particular circumstances to abandon all her possessions to the young female? I wish it had been in my power to solve this question otherwise than by mere probabilities, and that some misfortune had not befallen all the bees whose queen I had marked red on the thorax."

These expressions seem to indicate, that M. de Reaumur suspected that the old queens sometimes conducted the young swarms. By the following details, you will observe, that his suspicions are fully justified.

In the course of spring and summer, the same hive may throw several swarms. The old queen is always at the head of the first colony; the others are conducted by young queens. Such is the fact which I shall now prove; and the peculiarities attending it shall be related.

But previous to entering on this subject, I should repeat what has already been frequently observed, that the *leaf* or flat hives are indispensible in studying the industry and instinct of bees. When they are left at liberty to conduct several rows of parallel combs, we can no longer observe what is continually passing between them, or they must be dislodged by water or smoke, for examining what has been constructed; a violent proceeding, which has a material influence on their instinct, and consequently exposes an observer to the risk of supposing simple accidents permanent laws.

I now proceed to experiments proving that an old queen always conducts the first swarm.

One of my glass hives consisted of three parallel combs, placed in squares that opened like the leaves of a book. It was well peopled and abundantly provided with honey, wax, and brood, of every age. On the fifth of May 1788, I removed the queen, and on the sixth, transferred all the bees into another hive, with a fertile queen at least a year old. They entered easily and without fighting, and were in general well received. The old inhabitants of the hive, which, since privation of their queen, had begun twelve royal cells, also gave the fertile queen a good reception; they presented her with honey, and formed regular circles around her. However, there was a little agitation in the evening, but confined to the surface of the comb where we had put the queen, and which she had not quitted. All was perfectly quiet on the other side of this comb.

In the morning of the seventh, the bees had destroyed the twelve royal cells, but, independent of that, order continued prevalent in the hive; the queen laid the eggs of males in the large cells, and those of workers in the small ones, respectively.

Towards the twelfth, we found the bees occupied in constructing twenty-two royal cells, of the same species described by M. de Reaumur, that is the bases not in the plane of the comb, but appended perpendicularly by pedicles or stalks of different length, like stalactites, on the edge of the passage made by the bees through their combs. They bore considerable resemblance to the cup of an acorn, and the longest were only about two lines and a half in depth from the bottom to the orifice.

On the thirteenth, the queen seemed already more slender than when introduced into the hive; however she still laid some eggs, both in common cells and those of males. We also surprised her this day laying in a royal cell: she first dislodged the worker there employed, by pushing it away with her head, and then supported herself by the adjoining cells while depositing the egg.

On the fifteenth, the queen was still more slender: the bees continued their attention to the royal cells, which were all unequally advanced; some to three or four lines in height, while others were already an inch long; which proved that the queen had not laid in the whole at the same time.

At the moment when least expected, the hive swarmed on the nineteenth; we were warned of it by the noise in the air; and hastened to collect and put the bees into a hive purposely prepared. Though we had overlooked the facts attending the departure of the swarm, the object of this experiment was fulfilled; for, on examination of all the bees, we were convinced they had been conducted by the old queen; by that we introduced on the sixth of the month, and which had been deprived of one of the antennæ. Observe, there was no other queen in this colony. In the hive she had left, we found seven royal cells close at the top, but open at the side, and quite empty. Eleven more were sealed; and some others newly begun; no queen remained in the hive.

The new swarm next became the object of our attention: we observed it during the rest of the year, during winter and the subsequent spring; and, in April, we had the satisfaction of seeing a new swarm depart with the same queen at its head that had conducted the former swarm in May the preceding year.

You will remark, Sir, that this experiment is positive. We put an old queen in a glass hive while laying the eggs of males. The bees received her well, and at that time began to construct royal cells; she laid in one of them before us; and in the last place led forth the swarm.

We have several times repeated the same experiment with equal success. Thus it appears incontestible, that the old queen always conducts the first swarm; but never quits the hive before depositing eggs in the royal cells, from which other queens will proceed after her departure. The bees prepare these cells only while the queen lays male eggs; and a remarkable fact attends it, that after this laying terminates, her belly being considerably diminished, she can easily fly, whereas, her belly is previously so heavy she can hardly drag it along. Therefore it is necessary she should lay in order to be in a condition for undertaking her journey, which may sometimes be very long.

But this single condition is not enough. It is also requisite that the bees be very numerous: they should even be superabundant, and a person might say they are aware of it: for, if the hive is thin, no royal cells are constructed when the male eggs are laid, which is solely at the period that the queen is able to conduct a colony. This fact was proved by the following experiment on a large scale.

On the third of May 1788, we divided eighteen hives into two portions; all the queens were about a year old. Thus each portion of the hives had but half the bees that were originally there. Eighteen halves wanted queens, but the other eighteen had very fertile ones. They soon began to lay the eggs of males; but, the bees being few, they did not construct royal cells, and none of the hives threw a swarm.—Therefore, if the hive containing the old queen is not very populous, she remains in it until the subsequent spring; and if the population is then sufficient, royal cells will be constructed: the queen will begin to lay male eggs, and, after depositing them, will issue forth at the head of a colony, before the young queens are produced.

Such is a very brief abstract of my observations on swarms conducted by old queens. You must excuse the long detail on which I am about to enter, concerning the history of the royal cells left by the queen in the hive. Every thing relative to this part of the history of bees has hitherto been very obscure. A long course of observations, protracted even during several years, was necessary to remove, in some degree, the veil that concealed these mysteries. I have been indemnified for the trouble, indeed, by the pleasure of seeing my experiments reciprocally confirmed; but, considering the assiduity required in these researches, they were truly very laborious.

Having established in 1788 and 1789, that queens a year old conducted the first swarm, and that they left worms or nymphs in the hive to transform into queens in their turn; I endeavoured, in 1790, to profit by the goodness of the spring, to study all that related to these young queens; and I shall now extract the chief experiments from my journal.

On the fourteenth of May, we introduced two portions of bees, from the straw hives, into a large glass hive very flat; and allowed them only one queen of the preceding year, and which had already commenced laying in its native hive. We introduced her on the fifteenth. She was very fertile. The bees received her well, and she soon began to lay in large and small cells alternately.

On the twentieth, we saw the formation of twelve royal cells, all on the edges of the communications, or passages through the combs, and shaped liked stalactites.

On the twenty-seventh, ten were much but unequally enlarged; but none so long as when the worms are hatched.

On the twenty-eighth, previous to which the queen had not ceased laying, her belly was very slender, and she began to exhibit signs of agitation. Her motion soon became more lively, yet she still continued examining the cells as when about to lay; sometimes introducing half her belly, but suddenly withdrawing it, without having laid. At other times she deposited an egg, which lay in an irregular position, on one side of the hexagon, and not fixed by an end to the bottom of the cell. The queen produced no distinct sound in her course, and we heard nothing different from the ordinary humming of bees. She passed over those in her way; sometimes when she stopped, the bees meeting her also stopped; and seemed to consider her. They advanced briskly, struck her with their antennæ, and mounted on her back. She then went on carrying some of the workers on her back. None gave her honey, but she voluntarily took it from the cells in her way. The bees no longer inclosed and formed regular circles around her. The first, aroused by her motions, followed her running in the same manner, and in their passage excited those still tranquil on the combs. The way the queen had traversed was evident after she left it, by the agitation created, which was never afterwards quelled: she had soon visited every part of the hive, and occasioned a general agitation; if some places still remained tranquil, the bees in agitation arrived, and communicated their motion. The queen no longer deposited her eggs in cells; she let them fall fortuitously: nor did the bees any longer watch over the young; they ran about in every different direction; even those returning from the fields, before the agitation came to its height, no sooner entered the hive than they participated in these tumultuous motions. They neglected to free themselves of the waxen pellets on their limbs, and ran blindly about. At last the whole rushed precipitately towards the outlets of the hive, and the queen along with them.

As it was of much consequence to see the formation of new swarms in this hive, and, for that reason, as I wished it to continue very populous, I removed the queen, at the moment she came out, that the bees might not fly too far, and that they might return. In fact, after losing their female, they did return to the hive. To increase the population still more, I added another swarm, which had come from a straw hive on the same morning, and removed its queen also.

All these facts were certain, and appeared susceptible of no error. Notwithstanding this, I was particularly earnest to learn whether old queens always followed the same course; which induced me, on the twenty-ninth, to replace, in the glass hive, the queen a year old, which had hitherto been the subject of my experiments, and had just began to lay the eggs of males. On the same day, we found one of the royal cells left by the preceding queen larger than the rest; and, from its length, supposed the included worm two days old: the egg had, therefore, been laid on the twenty-fourth by that queen, and the worm was hatched on the twenty-seventh. On the thirtieth, the queen laid a great deal in the large and small cells alternately. Now, and the two following days, the bees enlarged several royal cells, but unequally, which proved that they included larvæ of different ages. One was closed on the first of June, and on the second another. The bees also commenced some new ones. All was perfectly quiet at eleven in the morning; but, at mid-day, the queen, from the utmost tranquillity, became evidently agitated; and her agitation insensibly communicated to the workers in every part of their dwelling. In a few minutes they precipitately crowded to the entrances, and, along with the queen, left the hive. After they had settled on the branch of a neighbouring tree, I sought for the queen; thinking that, by removing her, the bees might return to the hive, which actually ensued. Their first care seemed to consist in seeking the female; they were still in great agitation, but gradually calmed; and in three hours complete tranquillity was restored.

They had resumed their usual occupations on the third: they attended to the young, worked within the open royal cells, and also watched on those that were shut. They made a waved work on them, not by applying wax cordons, but by removing wax from the surface. Towards the top this waved work is almost imperceptible; it becomes deeper above, and the workers excavate it still more from thence to the base of the pyramid. The cell, when once shut, also becomes thinner; and is so much so, immediately preceding the queen's metamorphosis from a nymph, that all its motions are perceptible through the thin covering of wax on which the waved work is founded. It is a very remarkable circumstance, that in making the cells thinner, from the moment they are closed, the bees know to regulate their labour so that it terminates only when the nymph is ready to undergo its last metamorphosis.

On the seventh day the coccoon is almost completely *unwaxed*, if I may use the expression, at the part next to the head and thorax of the queen. This operation facilitates her exit; for she has nothing to do but cut the silk that forms the coccoon. Most probably the object is, to promote evaporation of the superabundant fluids of the nymph. I have made some direct experiments to ascertain the fact, but they are yet unfinished. A third royal cell was closed by the bees on the same day, the third of June, twenty-four hours after closing the second. The like was done to other royal cells successively, during the subsequent days.

Every moment of the seventh, we expected the queen to leave the royal cell shut on the thirtieth of May. The seven days had elapsed. The waving of her cell was so deep, that what passed within was pretty perceptible; we could discern that the silk of the coccoon was cut circularly, a line and a half from the extremity; but the bees being unwilling that she should yet quit her cell, they had soldered the covering to it with some particles of wax. What seemed most singular was, that this female emitted a very distinct sound, or clacking from her prison. It was still more audible in the evening, and even consisted of several monotonous notes in rapid succession.

The same sound proceeded from the royal cell on the eighth. Several bees kept guard round each royal cell.

The first cell opened on the ninth. The young queen was lively, slender, and of a brown colour. Now, we understood why bees retain the female captive in their cells, after the period for transformation has elapsed; it is, that they may be able to fly the instant they are hatched. The new queen occupied all our attention. When she approached the other royal cells, the bees on guard pulled, bit her, and chased her away; they seemed to be greatly irritated against her, and she enjoyed tranquillity only when at a good distance from these cells. This procedure was frequently repeated through the day. She twice emitted the sound; in doing so she stood, her thorax against a comb, and the wings crossed on her back; they were in motion but without being unfolded or further opened. Whatever might be the cause of her assuming this attitude, the bees were affected by it; all hung down their heads, and remained motionless.

The hive presented the same appearances on the following day. Twenty-three royal cells yet remained, assiduously guarded by a great many bees. When the queen approached, all the guards became agitated, surrounded her on all sides, bit, and commonly drove her away; sometimes when in these circumstances, she emitted her sound, assuming the position just described, from that moment the bees became motionless.

The queen confined in the second cell had not yet left it, and was heard to hum several times. We accidentally discovered how the bees fed her. On attentive examination, a small aperture was perceptible in the end of the coccoon which she had cut to escape, and which her guards had again covered with wax, to confine her still longer. She thrust her trunk through the cleft; at first the bees did not observe it alternately thrust out and drawn in, but one at length perceiving it, came to apply its trunk to that of the captive queen, and then gave way to others that also approached her with honey. When satisfied she retracted her trunk, and the bees again closed up the opening with wax.

The queen this day between twelve and one became extremely agitated. The royal cells had multiplied very much; she could go no where without meeting them, and on approaching she was very roughly treated. Then she fled, but to obtain no better reception. At last, these things agitated the bees; they precipitately rushed through the outlets of the hive, and settled on a tree in the garden. It singularly happened that the queen was herself unable to follow or conduct the swarm. She had attempted to pass between two royal cells before they were abandoned by the bees guarding them, and she was so confined and maltreated as to be incapable of moving. We then removed her into a separate hive prepared for a particular experiment; the bees, which had clustered on a branch, soon discovered their queen was not present, and returned of their own accord to the hive. Such is an account of the second colony of this hive.

We were extremely solicitous to ascertain what would become of the other royal cells. Four of the close ones had attained complete maturity, and the queens would have left them had not the bees prevented it. They were not open either previous to the agitation of the swarms, or at the moment of swarming.

None of the queens were at liberty on the eleventh. The second should have transformed on the eighth; thus she had been three days confined, a longer period than the first which formed the swarm. We could not discover what occasioned the difference in their captivity.

On the twelfth, the queen was at last liberated, as we found her in the hive. She had been treated exactly as her predecessor; the bees allowed her to rest in quiet, when distant from the royal cells, but tormented her cruelly when she approached them. We watched this queen a long time, but not aware that she would lead out a colony, we left the hive for a few hours. Returning at mid-day, we were greatly surprised to find it almost totally deserted. During our absence, it had thrown a prodigious swarm, which still clustered on the branch of a neighbouring tree. We also saw with astonishment the third cell open, and its top connected to it as by a hinge. In all probability the captive queen, profiting by the confusion that preceded the swarming, escaped. Thus, there was no doubt of both queens being in the swarm. We found it so; and removed them, that the bees might return to the hive, which they did very soon.

While we were occupied in this operation, the fourth captive queen left her prison, and the bees found her on returning. At first they were very much agitated, but calmed towards the evening, and resumed their wonted labours. They formed a strict guard around the royal cells, and took great care to remove the queen whenever she attempted to approach. Eighteen royal cells now remained to be guarded.

The fifth queen left her cell at ten at night; therefore two queens were now in the hive. They immediately began fighting, but came to disengage themselves from each other. However they fought several times during the night without any thing decisive. Next day, the thirteenth, we witnessed the death of one, which fell by the wounds of her enemy. This duel was quite similar to what is said of the combats of queens.

The victorious queen now presented a very singular spectacle. She approached a royal cell, and took this moment to utter the sound, and assume that posture, which strikes the bees motionless. For some minutes, we conceived, that taking advantage of the dread exhibited by the workers on guard, she would open it, and destroy the young female; also she prepared to mount the cell; but in doing so she ceased the sound, and quitted that attitude which paralyses the bees. The guardians of the cell instantly took courage; and, by means of tormenting and biting the queen, drove her away.

On the fourteenth, the sixth young queen appeared, and the hive threw a swarm, with all the concomitant disorder before described. The agitation was so considerable, that a sufficient number of bees did not remain to guard the royal cells, and several of the imprisoned queens were thus enabled to make their escape. Three were in the cluster formed by the swarm, and other three remained in the hive. We removed those that had led the colony, to force the bees to return. They entered the hive, resumed their post around the royal cells, and maltreated the queens when approaching.

A duel took place in the night of the fifteenth, in which one queen fell. We found her dead next morning before the hive; but three still remained, as one had been hatched during night. Next morning we saw a duel. Both combatants were extremely agitated, either with the desire of fighting, or the treatment of the bees, when they came near the royal cells. Their agitation quickly communicated to the rest of the bees, and at mid-day they departed impetuously with the two females. This was the fifth swarm that had left the hive between the thirtieth of May and fifteenth of June. On the sixteenth, a sixth swarm cast, which I shall give you no account of, as it shewed nothing new.

Unfortunately we lost this, which was a very strong swarm; the bees flew out of sight, and could never be found. The hive was now very thinly inhabited. Only the few bees that had not participated in the general agitation remained, and those that returned from the fields after the swarm had departed. The cells were, therefore, slenderly guarded; the queens escaped from them, and engaged in several combats, until the throne remained with the most successful.

Notwithstanding the victories of this queen, she was treated with great indifference from the sixteenth to the nineteenth, that is, the three days that she preserved her virginity. At length, having gone to seek the males, she returned with all the external signs of fecundation, and was henceforth received with every mark of respect; she laid her first eggs forty-six hours after fecundation.

Behold, Sir, a simple and faithful account of my observations on the formation of swarms. That the narrative might be the more connected, I have avoided interrupting it by the detail of several particular experiments which I made at the same time for elucidating various obscure points of their history. These shall be the subject of future letters. For, although I have said so much, I hope still to interest you.

PREGNY, *6. September 1791.*

P. S.—In revising this letter, I find I have neglected taking notice of an objection that may embarrass my readers, and which ought to be answered.

After the first five swarms had thrown, I had always returned the bees to the hive: it is not surprising, therefore, that it was continually so sufficiently stocked that each colony was numerous. But things are otherwise in the natural state: the bees composing a swarm do not return to the hive; and it will undoubtedly be asked, What resource enables a common hive to swarm three or four times without being too much weakened?

I cannot lessen the difficulty. I have observed that the agitation, which precedes the swarming, is often so considerable, that most of the bees quit the hive, and in that case we cannot well comprehend how, in three or four days afterwards, it can be in a state to send out another colony equally strong.

But remark, in the first place, that the queen leaves a prodigious quantity of workers' brood, which soon transforms to bees; and in this way the population sometimes becomes almost as great after swarming as before it.

Thus the hive is perfectly capable of affording a second colony without being too much weakened. The third and fourth swarm weaken it more sensibly; but the inhabitants always remain in sufficient numbers to preserve the course of their labours uninterrupted; and the losses are soon repaired by the great fecundity of the queen, as she lays above an hundred eggs a day.

If, in some cases, the agitation of swarming is so great, that all the bees participate in it, and leave the hive, the desertion lasts but for a moment. The hive throws only during the finest part of the day, and it is then that the bees are ranging through the country. Those that are out, therefore, cannot share in the agitation; when returned to the hive, they quietly resume their labours; and their number is not small, for, when the weather is fine, at least a third of the bees are employed in the fields at once.

Even in the most embarrassing case, namely, where the whole bees desert the hive, it does not follow, that all those endeavouring to depart become members of the new colony. When this agitation or delirium seizes them, the whole rush forward and accumulate towards the entrance of the hive, and are heated in such a manner that they perspire copiously. Those near the bottom, and supporting the weight of all the rest, seem drenched in perspiration; their wings grow moist; they are incapable of flight; and even when able to escape, they advance no farther than the board of the hive, and soon return.

Those that have lately left their cells remain behind the swarm, still feeble, they could not support themselves in flight. Here then are also many recruits to people what we should have thought a deserted habitation.

LETTER X.

THE SAME SUBJECT CONTINUED.

To preserve greater regularity in continuing the history of swarms, I think it proper to recapitulate in a few words the principal points of the preceding letter, and to expatiate on each, concerning the result of new experiments, respecting which I have still been silent.

In the first place. *If at the return of spring, we examine a hive well peopled, and governed by a fertile queen, we shall see her lay a prodigious number of male eggs in the course of May, and the workers will chuse that moment for constructing several royal cells of the kind described by M. de Reaumur.* Such is the result of several long continued observations, among which there has not been the slightest variation, and I cannot hesitate in announcing it as demonstrated. However, I should here add the necessary explanation. It is necessary that the queen, before commencing her *great* laying of the eggs of males, be eleven months old; when young she lays only those of workers. A queen, hatched in spring, will perhaps lay fifty or sixty eggs of drones in whole, but before beginning her great laying of them, which should be two thousand in a month, she must have completed her eleventh month in age. In the course of our experiments, which more or less disturbed the natural state of things, it often happened that the queen did not attain this age until October, and immediately began laying male eggs. The workers, as if induced by some emanation from the eggs, also adopted this time for building the royal cells. No swarm resulted thence, it is true, because in autumn all the necessary circumstances are absolutely wanting, but it is not less evident, that there is a secret relation between the production of the eggs of males, and the construction of royal cells.

This laying commonly continues thirty days. The bees on the twentieth or twenty-first lay the foundation of several royal cells. Sometimes they build sixteen or twenty; we have even had twenty-seven. When the cells are three or four lines high, the queen lays those eggs from which her own species will come, but not the whole in one day. That the hive may throw several swarms, it is essential that the young females conducting them be not all produced at the same time. One may affirm, that the queen anticipates the fact, for she takes care to allow at least the interval of a day between every egg deposited in the cells. It is proved by the bees knowing to close the cells the moment the worms are ready to metamorphose to nymphs. Now, as they close all the royal cells at different periods, it is evident the included worms are not all of an equal age.

The queen's belly is very turgid before she begins laying the eggs of drones; but it sensibly decreases as she advances, and when terminated is very small. Thus she finds herself in a condition to undertake a journey which circumstances may prolong; thus this condition was necessary; and as every thing is harmonious in the laws of nature, the origin of the males corresponds with that of the females, which they are to fecundate.

Secondly. *When the larvæ hatched from the eggs laid by the queen, in the royal cells, are ready to transform to nymphs, this queen leaves the swarm conducting a swarm along with her; and* *the first swarm that proceeds from the hive is uniformly conducted by the old queen*[*]. I think I can divine the reason of it.

That there may never be a plurality of females in a hive, nature has inspired queens with a natural horror against each other; they never meet without endeavouring to fight, and to accomplish their mutual destruction. Thus, the chance of combat is equal between them, and fortune will decide to which the empire shall pertain. But if one combatant is older than the rest, she is stronger, and the advantage will be with her. She will destroy her rivals successively as produced. Thus, if the old queen did not leave the hive, when the young ones undergo their last metamorphosis, it could produce no more swarms, and the species would perish. Therefore, to preserve the species, it is necessary that the old queen conduct the first swarm. But what is the secret means employed by nature to induce her departure? I am ignorant of it.

In this country it is very rare, though not without example, for the swarm, led forth by the old queen, in three weeks to produce a new colony, which is also conducted by the same old queen; and that may happen thus. Nature has not willed that the queen shall quit the first hive before her production of male eggs is finished. It is necessary for her to be freed of them, that she may become lighter. Besides, if her first occupation, on entering a new dwelling, was laying more male eggs, still she might perish either from age or accident before depositing those of workers. The bees in that case would have no means of replacing her, and the colony would go to ruin.

All these things have been with infinite wisdom foreseen. The first operation of the bees of a swarm is to construct the cells of workers. They labour at them with great ardour, and as the ovaries of the queen have been disposed with admirable

---

[*] Schirach seems to have been aware of this fact.—T.

foresight, the first eggs she has to lay in her new abode are those of workers. Commonly her laying continues ten or eleven days; and at this time portions of comb containing large cells are fabricated. It may be affirmed, that the bees know their queen will also lay the eggs of drones; she actually does begin to deposit some, though in much smaller number than at first; enough however to encourage the bees to construct royal cells. Now, if in these circumstances the weather is favourable, it is not impossible that a second colony may be formed, and that the queen may depart at the head of it within three weeks of conducting the first swarm. But I repeat, the fact is rare in our climate. Let me now return to the hives from which the queen has led the first colony.

Thirdly. *After the old queen has conducted the first swarm from the hive, the remaining bees take particular care of the royal cells, and prevent the young queens successively hatched from leaving them, unless at an interval of several days between each.*

In the preceding letter, I have given you the detail and proof of this fact, and I shall here add some reflexions. During the period of swarming, the conduct or instinct of bees seems to receive a particular modification. At all other times, when they have lost their queen, they appropriate workers worms to replace her; they prolong and enlarge the cells of these worms; they supply them with aliment more abundantly, and of a more pungent taste; and by this alteration, the worms that would have changed to common bees are transformed to queens. We have seen twenty-seven cells of this kind constructed at once; but when finished the bees no longer endeavour to preserve the young females from the attacks of their enemies. One may perhaps leave her cell, and attack all the other royal cells successively, which she will tear open to destroy her rivals, without the workers taking any part in their defence. Should several queens be hatched at once, they will pursue each other, and fight until the throne remain with her that is victorious. Far from opposing such duels, the other bees rather seem to excite the combatants.

Things are quite reversed during the period of swarming. The royal cells then constructed are of a different figure from the former. They resemble stalactites, and in the beginning are like the cup of an acorn. The bees assiduously guard the cells when the young queens are ready for their last metamorphosis. At length the female hatched from the first egg laid by the old queen leaves her cell; the workers at first treat her with indifference. But she, immediately yielding to the instinct which urges her to destroy her rivals, seeks the cells where they are enclosed; yet no sooner does she approach than the bees bite, pull, and drive her away, so that she is forced to remove; but the royal cells being numerous, scarce can she find a place of rest. Incessantly harassed with the desire of attacking the other queens, and incessantly repelled, she becomes agitated, and hastily traverses the different groups of workers, to which she communicates her agitation. At this moment numbers of bees rush towards the aperture of the hive, and, with the young queen at their head, depart to seek another habitation.

After the departure of the colony, the remaining workers set another queen at liberty, and treat her with equal indifference as the first. They drive her from the royal cells; being perpetually harassed, she becomes agitated; departs, and carries a new swarm along with her. In a populous hive this scene is repeated three or four times during spring. As the number of bees is so much reduced, that they are no longer capable of preserving a strict watch over the royal cells, several females then leave their confinement at once; they seek each other, fight, and the queen at last victorious reigns peaceably over the republic.

The longest intervals we have observed between the departure of each natural swarm have been from seven to nine days. This is the time that usually elapses after the first colony is led out by the old queen, until the next swarm is conducted by the first young queen set at liberty. The interval between the second and third is still shorter; and the fourth sometimes departs the day after the third. In hives left to themselves, fifteen or eighteen days are usually sufficient for the throwing of the four swarms, if the weather continues favourable, as I shall explain.

A swarm is never seen except in a fine day, or, to speak more correctly, at a time of the day when the sun shines, and the air is calm. Sometimes we have observed all the precursors of swarming, disorder and agitation, but a cloud passed before the sun, and tranquillity was restored; the bees thought no more of swarming. An hour afterwards, the sun having again appeared, the tumult was renewed; it rapidly augmented; and the swarm departed.

Bees generally seem much alarmed at the prospect of bad weather. While ranging in the fields the passage of a cloud before the hive induces them precipitately to return. I am induced to think they are disquieted by the sudden diminution of light. For if the sky is uniformly obscured, and there is no alteration in clearness or in the clouds dispelling, they proceed to the fields for their ordinary collections, and the first drops of a soft rain does not make them return with much precipitation.

I am persuaded that the necessity of a fine day for swarming is one reason that has induced nature to admit of bees protracting the captivity of their young queens in the royal cells. I will not deny that they sometimes seem to use this right in an arbitrary manner. However the confinement of the queens is always longer when bad weather lasts several days together. Here the final object cannot be mistaken. If the young females were at liberty to leave their cradles during these bad days, there would be a plurality of queens in the hive, consequently combats; and victims would fall. Bad weather might continue so long, that all the queens might at once have undergone their last metamorphosis, or attained their liberty. One victorious over the whole would enjoy the throne, and the hive, which should naturally produce several swarms, could give only one. Thus the multiplication of the species would have been left to the chance of rain, or fine weather, instead of which it is rendered independent of either, by the wise dispositions of nature. By allowing only a single female to escape at once, the formation of swarms is secured. This explanation appears so simple, that it is superfluous to insist farther on it.

But I should mention another important circumstance resulting from the captivity of queens; which is, that they are in a condition to fly, when the bees have given them liberty, and by this means are capable of profiting by the first moment of sunshine to depart at the head of a colony.

You well know, Sir, that all drones and workers are not in a condition to fly for a day or two after leaving their cells. Then they are of a whitish colour, weak, and their organs infirm. At least, twenty-four or thirty hours must elapse before the acquisition of perfect strength, and the development of all their faculties. It would be the same with the females was not their confinement protracted after the period of transformation; but we see them appear, strong, full grown, brown, and in a better condition for flying than at any other period. I have elsewhere observed, that constraint is used to retain the queens in captivity. The bees solder the covering to the sides of the cell by a cordon of wax. As I have also explained how they are fed, it need not be repeated here.

It is likewise a very remarkable fact, that queens are set at liberty earlier or later according to their age. Immediately when the royal cells were sealed, we marked them all with numbers, and we chose this period because it indicated the age of the queens exactly. The oldest was first liberated, then the one immediately younger, and so on with the rest. None of the younger queens were set at liberty before the older ones.

I have a thousand times asked myself how the bees could so accurately distinguish the age of their captives. Undoubtedly I should do better to answer this question by a simple avowal of my ignorance. At the same time, I must be permitted to state a conjecture. You will admit, that I have not, as some authors, abused the right of giving myself up to hypothesis; may not the humming or sound emitted by the young queens in their cells, be one of the methods employed by nature to instruct the bees in the age of their queens? It is certain that the female, whose cell is first sealed, is also the first to emit this sound. That in the next emits it sooner than the rest, and so on with those immediately subsequent. As their captivity may continue six days, it is possible that the bees in this space of time may forget which has emitted it first; but it is also possible, that the queens diversify the sounds, encreasing the loudness as they become older, and that the bees can distinguish these variations. We have even ourselves been able to distinguish differences in the sound, either with relation to the succession of notes, or their intensity; and probably there are gradations still more imperceptible that escape our organs, but may be sensible to those of the workers.

What gives weight to this conjecture is, that the queens brought up by M. Schirach's method, are perfectly mute; neither do the workers form any guard around their cells, nor do they retain them in captivity a moment beyond the period of transformation, and, when they have undergone it, they are allowed to combat until one has become victorious over all the rest. Why? Because the object is only to replace the last queen. Now, provided that among the worms reared as queens, only one succeeds, the fate of the others is uninteresting to the bees, whereas, during the period of swarming, it is necessary to preserve a succession of queens, for conducting the different colonies; and to ensure the safety of the queens, it is necessary to avert the consequences of the mutual horror by which they are animated against each other. Behold the evident cause of all the precautions that bees, instructed by nature, take during the period of swarming; behold an explanation of the captivity of females; and that the duration of their captivity might be ascertained by the age of the young queens, it was requisite for them to have some method of communicating to the workers when they should be liberated. This method consists in the sound emitted, and the variation they are able to give it.

In spite of all my researches, I have never been able to discover the situation of the organ which produces the sound. But I have instituted a new course of experiments on the subject, which are still unfinished.

Another problem still remains for solution. Why are the queens reared, according to M. Schirach's method, mute, whilst those bred in the time of swarming have the faculty of emitting a certain sound? What is the physical cause of this difference? At first I thought it might be ascribed to the period of life, when the worms that are to become queens receive the royal food. While hives swarm, the royal worms receive the food adapted for queens, from the moment of leaving the egg; those on the contrary, destined for queens, according to M. Schirach's method, receive it only the second or third day of their existence. It appears to me that this circumstance may have an influence on the different parts of organisation, and particularly on the organ of voice. Experiment has not confirmed this conjecture. I constructed glass cells in perfect imitation of royal cells, that the metamorphosis of the worms into nymphs, and of the nymphs to queens, might be visible. These experiments are related in a preceding letter. Into one of these artificial cells we introduced the nymph of a worm, reared according to M. Schirach's method, twenty-four hours before it could naturally undergo its last metamorphosis; and we replaced the glass cell in the hive, that the nymph might have the necessary degree of heat. Next day, we had the satisfaction of seeing it divest itself of the spoil, and assume its ultimate figure. This queen was prevented from escaping from her prison; but we had contrived an aperture for her thrusting out her trunk, and that the bees might feed her. I expected that she would have been completely mute; but it was otherwise; for she emitted sounds similar to those already described, therefore my conjecture was erroneous.

I next conceived that the queen being restrained in her motions, and in her desire for liberty, was induced to emit certain sounds. All queens, in this new point of view, are equally capable of emitting the sound, but to induce them to it, they must be in a confined situation. In the natural state, the queens that come from workers are not a single instant in restraint; and, if they do not emit the sound, it is because nothing impels them to it. On the other hand, those produced at the time of swarming are induced to do so by the captivity in which they are kept. For my own part, I give little weight to this conjecture; and though I state it here, it is less with a view to claim merit than to put others on a plan of discovering something more probable.

I do not ascribe to myself the credit of having discovered the humming of the queen bee. Old authors speak of it. M. de Reaumur cites a Latin work published 1671, *Monarchia Femina*, by Charles Butler. He gives a very brief abstract of this naturalist's observations, who we easily see has exaggerated or rather disguised the truth, by mixing it with the most absurd fancies; but it is not the less evident that Butler has heard this peculiar humming of queens, and that he did not confound it with the confused humming sometimes heard in hives.

Fourthly. *The young queens conducting swarms from their native hive are still in a virgin state.* The day after, being settled in their new abode, they generally depart in quest of the males; and this is usually the fifth day of their existence as queens; for two or three pass in captivity, one in their native hive, and a fifth in their new dwelling. Those queens that come from the worm of a worker, also pass five days in the hive before going in quest of males. So long as in a state of virginity, both are treated with indifference by the bees; but after returning with the external marks of fecundation, they are received by their subjects with the most distinguished respect. However, forty-six hours elapse after fecundation before they begin to lay. The old queen, which leads the first swarm in spring, requires no farther commerce with the males, for preservation of her fecundity. A single copulation is sufficient to impregnate all the eggs she will lay for at least two years.

PREGNY, 8. September 1791.

LETTER XI.

THE SAME SUBJECT CONTINUED.

I have collected my chief observations on swarms in the two preceding letters; those most frequently repeated, and of which the uniformity of result leads me to apprehend no error. I have deduced what seem the most direct consequences; and in all the theoretical part, I have sedulously avoided going beyond facts. What is yet to be mentioned is more hypothetical, but it engrosses several curious experiments.

It has been demonstrated, that the principal motive of the young females departing when hives swarm, is their insuperable antipathy to each other. I have repeatedly observed that they cannot gratify their aversion, because the workers with the utmost care prevent them from attacking the royal cells. This perpetual opposition at length creates a visible inquietude, and excites a degree of agitation that induces them to depart. All the young queens are successively treated alike in hives that are to swarm. But the conduct of the bees towards the old queen, destined to conduct the first swarm, is very different. Always accustomed to respect fertile queens, they do not forget what they owe to her; they allow her the most uncontrolled liberty. She is permitted to approach the royal cells; and if she even attempts to destroy them, no opposition is presented by the bees. Thus her inclinations are not obstructed, and we cannot ascribe her flight, as that of the young queens, to the opposition she suffers. Therefore, I candidly confess myself ignorant of the motives of her departure.

Yet, on more mature reflection, it does not appear to me that this fact affords so strong an objection against the general rule as I had at first conceived. It is certain at least, that the old queens, as well as the young ones, have the greatest aversion to the individuals of their own sex. This has been proved by the numerous royal cells destroyed. You will remember, Sir, that in my first observations on the departure of old queens, seven royal cells opened at one side were destroyed by the queen. If rain continues several days, the whole are destroyed; in this case, there is no swarm, which too often happens in our climate, where spring is generally rainy. Queens never attack cells containing an egg or a very young worm; but only when the worm is ready for transforming to a nymph, or when it has undergone its last metamorphosis.

The presence of royal cells with nymphs or worms near their change, also inspires old queens with the utmost horror or aversion; but here it would be necessary to explain why the queen does not always destroy them though it is in her power. On this point, I am limited to conjectures. Perhaps the great number of royal cells in a hive at once, and the labour of opening the whole, creates insuperable alarm in the old queen. She commences indeed with attacking her rivals; but, incapable of immediate success, her inquietude during this work becomes a terrible agitation. If the weather continues favourable, while she remains in this condition, she is naturally disposed to depart.

It may easily be understood, that the workers accustomed to respect their queen, whose presence is a real necessity to them, crowd after her; and the formation of the first swarm creates no difficulty in this respect. But you will undoubtedly ask, Sir, What motive can induce the workers to follow their queen from the hive, while they treat the young queens very ill, and, even in their most amicable moments, testify perfect indifference towards them. Probably it is to escape the heat to which the hive is then exposed. The extreme agitation of the females leads them to traverse the combs in all directions. They pass through groups of bees, injure and derange them; they communicate a kind of delirium, and these tumultuous motions raise the temperature to an insupportable degree. We have frequently proved it by the thermometer. In a populous hive it commonly stands between 92° and 97°, in a fine day of spring; but during the tumult which precedes swarming, it rises above 104°. And this is heat intolerable to bees. When exposed to it, they rush impetuously towards the outlets of the hive and depart. In general they cannot endure the sudden augmentation of heat, and in that case quit their dwelling; neither do those returning from the fields enter when the temperature is extraordinary.

I am certain, from direct experiments, that the impetuous courses of the queen over the combs, actually throws the workers into agitation; and I was able to ascertain it in the following manner. I wished to avoid a complication of causes. It was particularly important to learn, whether the queen would impart her agitation but not at the time of swarming. Therefore I took two females still virgins, but capable of fecundation for above five days, and put one in a glass hive sufficiently populous; the other I put into a different hive of the same kind. Then I shut the hives in such a way that there was no possibility of their escape. The air had free circulation. I then prepared to observe the hives every moment that the fineness of the weather would invite both males and females to go abroad, for the purpose of fecundation. Next morning, the weather being gloomy, no male left the hive, and the bees were tranquil; but towards eleven of the following day, the sun shining bright, both queens began to run about seeking an exit from every part of their dwelling; and from their inability to find one, traversed the combs with the most evident symptoms of disquiet and agitation. The bees soon participated of the same disorder; they crowded towards that part of the hive where the openings were placed; unable to escape they ascended with equal rapidity, and ran heedlessly over the cells until four in the afternoon. It is nearly about this time that the sun declining in the horizon recalls the males; queens requiring fecundation never remain later abroad. The two females became calmer, and tranquillity was in a short time restored. This was repeated several subsequent days with perfect similarity; and I am now convinced that there is nothing singular in the agitation of bees while swarming, but that they are always in a tumultuous state when the queen herself is in agitation.

I have but one fact more to mention. It has already been observed, that on losing the female, bees give the larvæ of simple workers the royal treatment, and, according to M. Schirach, in five or six days they repair the loss of their queen. In this case there are no swarms. All the females leave their cells almost at the same moment, and after a bloody combat the throne remains with the most fortunate.

I can very well comprehend that the object of nature is to replace the lost queen; but as bees are at liberty to choose either the eggs or worms of workers, during the first three days of existence; to supply her place, why do they give the royal treatment to worms, all of nearly an equal age, and which must undergo their last metamorphosis almost at the same time? Since they are enabled to retain the young females in their cells, why do they allow all the queens, reared according to Schirach's method, to escape at once. By prolonging their captivity more or less, they would fulfil two most important objects at once, in repairing the loss of their females and preserving a succession of queens to conduct several swarms.

At first it was my opinion, that this difference of conduct proceeded from the difference of circumstances in which they found themselves situated. They are induced to make all their dispositions relative to swarming only when in great numbers, and when they have a queen occupied with her principal laying of male eggs; whereas, having lost their female, the eggs of drones are no longer in the combs to influence their instinct. They are in a certain degree restless and discouraged.

Therefore, after removing the queen from a hive, I thought of rendering all the other circumstances as similar as possible to the situation of bees preparing to swarm. By introducing a great many workers, I encreased the population to excess, and supplied them with combs of male brood in every stage. Their first occupation was to construct royal cells after Schirach's method, and to rear common worms with royal food. They also began some stalactite cells, as if the presence of the male brood had inspired them to it; but this they discontinued, as there was no queen to deposit her eggs. Finally, I gave them several close royal cells, taken indifferently from hives preparing to swarm. However, all these precautions were fruitless; the bees were occupied only with replacing their lost queen; they neglected the royal cells entrusted to their care; the included queens came out at the ordinary time, without being detained prisoners a moment; they engaged in several combats, and there were no swarms.

Recurring to subtleties, we may perhaps suggest a cause for this apparent contradiction. But the more we admire the wise dispositions of the author of nature, in the laws he has prescribed to the industry of animals, the greater reserve is necessary in admitting any theory adverse to this beautiful system, and the more must we distrust that facility of imagination from which we think by embellishment to elucidate facts.

In general, Naturalists who have long observed animals, and those in particular who have chose insects for their favourite study, have too readily ascribed to them our sentiments, our passions, and even our intentions and designs. Incited to admiration, and disgusted perhaps by the contempt with which insects are treated, they have conceived themselves obliged to justify the consumption of time bestowed on this pursuit, and they have painted different traits of the industry of these minute animals, with the colours inspired by an exalted imagination. Nor is even the celebrated Reaumur to be acquitted of such a charge. He frequently ascribes combined intentions to bees; love, anticipation, and other faculties of too elevated a kind. I think I can perceive that although he formed very just ideas of their operations, he would be well pleased that his reader should admit they were sensible of their own interests. He is a painter who by a happy interest flatters the original, whose features he depicts. On the other hand, Buffon unjustly considers bees as mere automatons. It was reserved for you, Sir, to establish the theory of animal industry on the most philosophical principles, and to demonstrate that those actions that have a moral appearance depend on an association of ideas *simply sensible*. It is not my object here to penetrate those depths, or to insist on the details.

But, on the whole, facts relative to the formation of swarms perhaps present more subjects of admiration than any other part of the history of bees. I think it proper to state, in a few words, the simplicity of the methods by which the wisdom of nature guides their instinct. It cannot allow them the slightest portion of understanding; it leaves them no precautions to be taken, no combination to be followed, no foresight to exercise, no knowledge to acquire. But having adapted their sensorium to the different operations with which they are charged, it is the impulse of pleasure which leads them on. She has therefore pre-ordained all that is relative to the succession of their different labours; and to each operation she has united an agreeable sensation. Thus, when bees construct cells, watch over their larvæ, and collect provisions, we must not seek for method, affection, or foresight. The only inducement must be sought for in some pleasing sensation attached to each of these operations. I address a philosopher; and as these are his own opinions applied

118

to new facts, I believe my language will be easily understood. But I request my readers to peruse and to reflect on that part of your works which treats of the industry of animals. Let me add but another sentence. The inducement of pleasure is not the sole agent; there is another principle, the prodigious influence of which, at least with regard to bees, has hitherto been unknown, that is the sentiment of aversion which all females continually feel against each other, a sentiment whose existence is so fully demonstrated by my experiments, and which explains many important facts in the theory of swarms.

PREGNY, 10. September 1791.

LETTER XII.

ADDITIONAL OBSERVATIONS ON QUEENS THAT LAY ONLY THE EGGS OF DRONES, AND ON THOSE DEPRIVED OF THE ANTENNÆ.

In relating my first observations on queens that lay male eggs alone, I have proved that they lay them in cells of all dimensions indifferently, and even in royal cells. It is also proved that the same treatment is given to male worms hatched from eggs laid in the royal cells, as if they were actually to be transformed to queens; and I have added, that in this instance the instinct of the workers appeared defective. It is indeed most singular, that bees which know the worms of males so well when the eggs are laid in small cells, and never fail to give them a convex covering when about to transform to nymphs, should no longer recognise the same species of worms when the eggs are laid in royal cells, and treat them exactly as if they should change to queens. This irregularity depends on something I cannot comprehend.

In revising what is said on this subject, I observe still wanting an interesting experiment to complete the history of queens that lay only the eggs of drones. I had to investigate whether these females could themselves distinguish that the eggs they deposit in the royal cells would not produce queens. I have already observed that they do not endeavour to destroy these cells when close, and I thence concluded, that in general the presence of royal cells in their hive does not inspire them with the same aversion to females whose fecundation has been retarded; but to ascertain the fact more correctly, it was essential to examine how the presence of a cell containing a royal nymph would affect a queen that had never laid any other than the eggs of drones.

This experiment was easy; and I put it in practice on the fourth of September, in a hive some time deprived of its queen. The bees had not failed to construct several royal cells for replacing their females. I chose this opportunity for supplying them with a queen, whose fecundation had been retarded to the twenty-eighth day, and which laid none but the eggs of males. At the same time, I removed all the royal cells, except one that had been sealed five days. One remaining was enough to shew the impression it would make on the stranger queen introduced; had she endeavoured to destroy it; this, in my opinion, would have proved that she anticipated the origin of a dangerous rival. You must admit the use I make of the word anticipate; it saves a long paraphrase; I feel the impropriety of it. If, on the contrary, she did not attack the cell I would thence conclude that the delay of fecundation, which deprived her of the power of laying workers eggs, had also impaired her instinct. This was the fact; the queen passed several times over the royal cell, both the first and the subsequent day, without seeming to distinguish it from the rest. She quietly laid in the surrounding cells; notwithstanding the cares incessantly bestowed by the bees upon it, she never one moment appeared to suspect the danger with which the included royal nymph threatened her. Besides, the workers treated their new queen as well as they would have treated any other female. They were lavish of honey and respect, and formed those regular circles around her that seem an expression of homage.

Thus, independent of the derangement occasioned by retarded impregnation, in the sexual organs of queens, it certainly impairs their instinct. Aversion or jealousy is no longer preserved against their own sex in the nymphine state, nor do they longer endeavour to destroy them in their cradles.

My readers will be surprised that queens whose fecundation has been retarded, and whose fecundity is so useless to bees, should be so well treated and become as dear to them as females laying both kinds of eggs. But I remember to have observed a fact more astonishing still. I have seen workers bestow every attention on a queen though sterile; and after her death treat her dead body as they had treated herself when alive, and long prefer this inanimate body to the most fertile queens I had offered them. This sentiment, which assumes the appearance of so lively an affection, is probably the effect of some agreeable sensation communicated to bees by their queen, independent of fertility. Those laying only the eggs of males probably excite the same sensation in the workers.

I now recollect that the celebrated Swammerdam somewhere observes, that when a queen is blind, sterile, or mutilated, she ceases to lay, and the workers of her hive no longer labour or make any collections, as if aware that it was now useless to work. He cites no experiment that led him to the discovery. Those made by myself have afforded some very singular results.

I frequently amputated the four wings of queens; and not only did they continue laying, but the same confederation of them was testified by the workers as before. Therefore, Swammerdam has no foundation for asserting, that mutilated queens cease to lay. Indeed, from his ignorance of fecundation taking place without the hives, it is possible he cut the wings off virgin queens, and they, becoming incapable of flight, remained sterile from inability to seek the males in the air. Thus, amputation of the wings does not produce sterility in queens.

◆　　◆　　◆　　◆　　◆

I have frequently cut off one antennæ to recognise a queen the more easily, and it was not prejudicial to her either in fecundity or instinct nor did it affect the attention paid to her by the bees. It is true, that as one still remained, the mutilation was imperfect; and the experiment decided nothing. But amputation of both antennæ produced most singular effects. On the fifth of September, I cut both off a queen that laid the eggs of males only, and put her into the hive immediately after the operation. From this moment there was a great alteration in her conduct. She traversed the combs with extraordinary vivacity. Scarcely had the workers time to separate and recede before her; she dropped her eggs, without attending to deposit them in any cell. The hive not being very populous, part was without comb. Hither she seemed particularly earnest to repair, and long remained motionless. She appeared to avoid the bees; however, several workers followed her into this solitude, and treated her with the most evident respect. She seldom required honey from them, but, when that occurred, directed her trunk with an uncertain kind of feeling, sometimes on the head and sometimes on the limbs of the workers, and if it did reach their mouths, it was by chance. At other times she returned upon the combs, then quitted them to traverse the glass sides of the hive: and always dropped eggs during her various motions. Sometimes she appeared tormented with the desire of leaving her habitation. She rushed towards the opening, and entered the glass tube adapted there; but the external orifice being too small, after fruitless exertion, she returned. Notwithstanding these symptoms of delirium, the bees did not cease to render her the same attention as they ever pay to their queens, but this one received it with indifference. All that I describe appeared to me the consequence of amputating the antennæ. However, her organization having already suffered from retarded fecundation, and as I had observed her instinct in some degree impaired, both causes might possibly concur in producing the same effect. To distinguish properly what belonged to the privation of the antennæ, a repetition of the experiment was necessary, in a queen otherwise well organised, and capable of laying both kinds of eggs.

This I did on the sixth of September. I amputated both the antennæ of a female which had been several months the subject of observation, and being of great fecundity had already laid a considerable number of workers eggs, and those of males. I put her into the same hive where the queen of the preceding experiment still remained, and she exhibited precisely the same marks of delirium and agitation, which I think it needless to repeat. I shall only add, that to judge better of the effect produced by privation of the antennæ, on the industry and instinct of bees, I attentively considered the manner in which these two mutilated queens treated each other. You cannot have forgot, Sir, the animosity with which queens, possessing all their organs, combat, on which account it became extremely interesting to learn whether they would experience the same reciprocal aversion after losing their antennæ. We studied these queens a long time; they met several times in their courses, and without exhibiting the smallest resentment. This last instance is, in my opinion, the most complete evidence of a change operated in their instinct.

Another very remarkable circumstance, which this experiment gave me occasion to observe, consists in the good reception given by the bees to the stranger queen, while they still preserved the first. Having so often seen the symptoms of discontent that a plurality of queens occasions, after having witnessed the clusters formed around these supernumerary queens to confine them, I could not expect they would pay the same homage to a second mutilated one they still testified towards the first. Is it because after losing the antennæ, these queens have no more any characteristic which distinguishes the one from the other?

I was the more inclined to admit this conjecture from the bad reception of a third fertile queen preserving her antennæ, which was introduced into the same hive. The bees seized, bit her, and confined her so closely, that she could hardly breath or move. Therefore, if they treat two females deprived of antennæ in the same hive equally well, it is probably because they experience the same sensation from these two females, and want the means of longer distinguishing them from each other.

From all this, I conclude, that the antennæ are not a frivolous ornament to insects, but, according to all appearance, are the organs of touch or smell. Yet I cannot affirm which of these senses reside in them. It is not impossible that they are organised in such a manner as to fulfil both functions at once.

As in the course of this experiment both mutilated females constantly endeavoured to escape from the hive, I wished to see what they would do if set at liberty, and whether the bees would accompany them in their flight. Therefore I removed the first and third queen from the hive, leaving the fertile mutilated one, and enlarged the entrance.

The queen left her habitation the same day. At first she tried to fly, but, her belly being full of eggs, she fell down and never attempted it again. No workers accompanied her. Why, after rendering the queen so much attention while she lived among them, did they abandon her now on her departure? You know, Sir, that queens governing a weak swarm are sometimes discouraged, and fly away, carrying all their little colony along with them. In like manner sterile queens, and those whose dwelling is ravaged by weevils, depart; and are followed by all their bees. Why therefore in this experiment did the workers allow their mutilated queen to depart alone? All that I can hazard on this question is a conjecture. It appears that bees are induced to quit the hives from the increased heat which occasions the agitation of their queen, and the tumultuous motion which she communicates to them. A mutilated queen, notwithstanding her delirium, does not agitate the workers, because she seeks the uninhabited parts of the hive, and the glass panes of it: she hurries over clusters of bees, but the shock resembles that of any other body, and produces only a local and instantaneous motion. The agitation arising from it, is not communicated from one place to another, like that produced by a queen, which in the natural state wishes to abandon her hive and lead out a swarm; there is no increased heat, consequently nothing that renders the hive insupportable to her.

This conjecture, which affords a tolerable explanation why bees persist in remaining in the hive, though the mutilated queen has left it, is no reason for the motive inducing the queen herself to depart. Her instinct is altered; that is the whole that I can perceive. I can discern nothing more. It is very fortunate for the hive, that this queen departs, for the bees incessantly attend her; nor do they ever think of procuring another while she remains; and if she was long of leaving them, it would be impossible to replace her; for the workers worms would exceed the term at which they are convertible into royal worms, and the hive would perish. Observe, that the eggs dropped by the mutilated queen can never serve for replacing her, for, not being deposited in cells, they dry and produce nothing.

I have yet to say a few words on females laying male eggs only. M. Schirach supposes that one branch of their double ovary suffers some alteration. He seems to think that one of these branches contains the eggs of males, while the other has none but common eggs, and as he ascribes the inability of certain queens to lay the latter to some disease, his conjecture becomes very plausible. In fact, if the eggs of males and workers are indiscriminately mixed in both branches of the ovary, it appears at first sight that whatever cause acts on that organ, it should equally affect both species of eggs. If on the contrary, one branch is occupied by the eggs of drones only, and the other contains none but common eggs, we may conceive how disease affects the one, while the other remains untouched. Though this conjecture is probable, it is confuted by observation. We lately dissected queens, which laid none but male eggs, and found both branches of the ovary equally well expanded, and equally sound, if I may use the expression. The only difference that struck us was that in these two branches, the eggs were apparently not so close together as in the ovaries of queens laying both kinds of eggs.

PREGNY, 12. September 1791.

LETTER XIII.

ECONOMICAL CONSIDERATIONS ON BEES.

In this letter I shall treat of the advantages that may be derived from the new invented hives, called *book* or *leaf* hives, in promoting the *economical knowledge* of bees.

It is needless to relate the different methods hitherto employed in forcing bees to yield up a portion of their honey and wax; all resemble each other in being cruel and ill understood.

It is evident, when bees are cultivated for the purpose of sharing the produce of their labours, we must endeavour to multiply them as much as the nature of the country admits; and consequently to regard their lives at the time we plunder them. Therefore it is an absurd custom to sacrifice whole hives to get at the riches they contain. The inhabitants of this country, who follow no other method, annually lose immense numbers of hives; and spring, being generally unfavourable to swarms, the loss is irreparable. I well know that at first they will not adopt any other method; they are too much attached to prejudices and old customs. But naturalists and intelligent cultivators of bees will be sensible of the utility of the method I propose; and if they apply it to use I hope their example will extend and perfect the culture of bees.

It is not more difficult to lodge a natural swarm in a leaf hive than in any other of a different shape. But there is one precaution essential to success, which I should not omit. Though the bees are indifferent as to the position of their combs, and as to their greater or lesser size, they are obliged to construct them perpendicular to the horizon, and parallel to each other. Therefore, if left entirely to themselves, when establishing a colony in one of those new hives, they would frequently construct several small combs parallel indeed, but perpendicular to the plane of the frames or leaves, and by this disposition prevent the advantages which I think to derive from the figure of my hives, since they could not be opened without breaking the combs. Thus they must previously have a guide to follow; the cultivator himself lays the foundation of their edifices, and that by a simple method. A portion of comb must be solidly fixed in some of the boxes composing the hive; the bees will extend it; and, in prosecution of their work, will accurately follow the plan already given them. Therefore on opening the hive, no obstacle is to be removed, nor stings to be dreaded, for one of the most singular and valuable properties attending this construction, is its rendering the bees tractable. I appeal to you, Sir, for the truth of what I say. In your presence I have opened all the divisions of the most populous hives, and the tranquillity of the bees has given you great surprise. I can desire no other evidence of my assertion. It is in the facility of opening these hives at pleasure that all the advantages lie, which I expect in perfecting the economical knowledge of bees.

I conceive, when I observe bees may be rendered tractable, that it need not be added, I do not arrogate to myself the absurd pretence of *taming* them, for this excites a vague idea of tricks; and I would willingly avoid the hazard of exposing myself to any such reproach. I ascribe their tranquillity on opening the hives, to the manner that the sudden introduction of light affects them; then, they seem rather to testify fear than anger. Many retire and enter the cells, and appear to conceal themselves. What confirms my conjecture is, their being less tractable during night or after sunset than through the day. Thus, we must open the hives, while the sun is above the horizon, cautiously, and without any sudden shock. The divisions must be separated slowly, and care taken not to wound any of the bees. If they cluster too much on the combs, they must be brushed off with a feather; and breathing on them carefully avoided. The air we expire seems to excite their fury; it certainly has some irritating quality, for if bellows are used, they are rather disposed to escape than to sting.

Respecting the advantages of leaf hives, I shall observe, they are very convenient for forming *artificial* swarms. In the history of natural swarms, I have shewn how many favourable circumstances are necessary for their success. From experience I know that they very often fail in our climate; and even when a hive is disposed to swarm, it frequently happens that the swarm is lost either because the moment of its departure has not been foreseen, because it rises out of sight, or settles on inaccessible places. Instructing the cultivators of bees how to make artificial swarms is a real service, and the form of my hives renders this an easy operation. But it requires farther illustration.

Since bees, according to M. Schirach's discovery, can procure another queen after having lost their own, provided there is workers brood in the combs not above three days old, it results that we can at pleasure produce queens, by removing the reigning one. Therefore, if a hive sufficiently populous is divided in two, one half will retain the old queen, and the other will not be long of obtaining a new one. But to ensure success, we must choose a propitious moment, which is never certain but in leaf hives. In these we can see whether the population is sufficient to admit of division, if the brood is of the proper age, if males exist or are ready to be produced for impregnating the young queens.

Supposing the union of all these conditions, the following is the method to be pursued. The leaf hive may be divided through the middle without any shock. Two empty divisions may be insinuated between the halves, which, when exactly applied to each other, are close on the outside. The queen must be sought in one of the halves, and marked to avoid mistake. If she by chance remains in the division with most brood, she is to be transferred to the other with less, that the bees may have every possible opportunity of obtaining another female. Next, it is necessary to connect the halves together, by a cord tied tight around them, and care must be taken that they are set on the same board that the hive previously occupied. The old entrance, now become useless, will be shut up; but as each half requires a new one, it ought to be made at the bottom of each division, on purpose that they may be as far asunder as possible. Both entrances should not be made on the same day. The bees in the half deprived of the queen ought to be confined twenty-four hours, and no opening made before then except for admission of air. Without this precaution, they would soon search for their queen, and infallibly find her in the other division. They will then retire in great numbers from their own division, until too few remain to perform the necessary labours. But this will not ensue if they are confined twenty-four hours, provided that interval is sufficient to make them forget the queen. When all these circumstances are favourable, the bees, in the division wanting the queen, will the same day begin to labour in procuring another, and ten or fifteen days after the operation, their loss will be repaired. The young female they have reared, soon issues forth to seek impregnation, and in two days commences the laying of workers eggs. Nothing more is wanting to the bees of this half hive, and the success of the artificial swarm is ensured.

It is to M. Schirach that we are indebted for this ingenious method of forming swarms. He supposes, by producing young queens early in spring, that early swarms might be procured, which would certainly be advantageous in favourable circumstances. But unfortunately this is impossible. Schirach believed that queens were impregnated of themselves, consequently he thought that after being artificially produced, they would lay and give birth to a numerous posterity. Now, this is an error; the females, to become fertile, require the concourse of the males, and if not impregnated within a few days of their origin, their laying, as I have observed, is completely deranged. Thus, if a swarm were artificially formed before the usual time of the males originating, the bees would be discouraged by the sterility of the young female. Or should they remain faithful to her, awaiting the period of fecundation, as she could not for three or four weeks receive the approaches of the male, she would lay eggs producing males only, and the hive in this case would perish. Thus the natural order must not be deranged, but we must delay the division of hives until males are about to originate or actually exist.

Besides, if M. Schirach did succeed in obtaining artificial swarms, notwithstanding the great inconvenience of his hives, it was owing to his singular address and unremitting assiduity. He had some pupils in the art; these communicated the method of forming artificial swarms to others, and there are people now in Saxony who traverse the country practising this operation. Those versant in the matter can alone venture to undertake it with common hives, whereas, every cultivator can do it himself with the leaf hives.

In this construction, another very great advantage will also be found. Bees can be forced to work in wax. Here I am led to what I believe is a new observation. While naturalists have directed our admiration to the parallel position of the combs, they have overlooked another trait in the industry of bees, namely, the equal distance uniformly between them. On measuring the interval separating the combs, it will generally be found four lines. Were they too distant, it is very evident the bees would be much dispersed and unable to communicate their heat reciprocally; whence the brood would not be exposed to sufficient warmth. Were the combs too close, on the contrary, the bees could not freely traverse the intervals, and the work of the hive would suffer. Therefore, a certain distance always uniform is requisite, which corresponds equally well with the service of the hive, and the care necessary for the worms. Nature, which has taught bees so much, has instructed them regularly to preserve this distance. At the approach of winter, they sometimes elongate the cells which are to contain the honey, and thus contract the intervals between the combs. But this operation is a preparation for a season, when it is important to have plentiful magazines, and when their activity being very much relaxed, it is unnecessary for their communications to be so spacious and free. On the return of spring, the bees hasten to contract these elongated cells, that they may become fit for receiving the eggs which the queen will lay, and thus re-establish the just distance which nature has ordained.

This being admitted, bees may be forced to work in wax, or, which is the same thing, to construct new combs. To accomplish the object, it is only necessary to separate those already built so far asunder that they may build others in the interval. Suppose an artificial swarm is lodged in a leaf hive, composed of six divisions, each containing a comb, if the young queen is as fertile as she ought to be, the bees will be very active in their labours, and disposed to make great collections of wax. To induce them towards this an empty box or division must be placed between two others, each containing a comb. As all the boxes are of equal dimensions, and of the necessary width for receiving a comb, the bees having sufficient space for constructing a new one in the empty division introduced into the hive, will not fail to build it, because they are under the necessity of never having more than four lines between them. Without any guide, this new comb will be parallel to the old ones, to preserve that law which establishes an equal distance throughout the whole.

If the hive is strong and the weather good, three empty divisions may at first be left between the old combs; one between the first and second, another between the third and fourth, and the last between the fifth and sixth. The bees will fill them in seven or eight days, and the hive then contains nine combs. Should the temperature of the weather continue favourable, three new leaves or divisions may be introduced; consequently in fifteen days or three weeks, the bees will have been forced to construct six new combs. The experiment may be extended farther in warm climates, and where flowers perpetually blow. But in our country, I have reason to think that the labour should not be forced more during the first year.

From these details, you are sensible, Sir, how preferable *leaf hives* are to those of any other construction, and even to those ingenious stages described by *M. Palteau*, for the bees cannot by means of them be forced to labour more in wax than they would do if left to themselves; whereas, they are obliged to do it by inserting empty divisions. Next, the combs constructed on those stages cannot be removed without destroying considerable portions of brood, deranging the bees, and creating real disorder in the hive.

Mine have also this advantage, that what passes within may daily be observed, and we may judge of the most favourable moments for depriving the bees of part of their stores. With all the combs before us we can distinguish those containing brood only, and what it is proper to preserve. The scarcity or abundance of provisions is visible, and the portion suitable may be taken away.

I should protract this letter too much, if I gave an account of all my observations on the time proper for inspecting hives, on the rules to be followed in the different seasons, and the proportion to be observed in dividing their riches with them. The subject would require a separate work; and I may perhaps one day engage in it; but until that arrives I shall always feel gratification in communicating to cultivators, who wish to follow my method, directions of which long practice has demonstrated the utility.

Here I shall only observe, that we hazard absolute ruin of the hives, by robbing them of too great a proportion of honey and wax. In my opinion, the art of cultivating these animals consists in moderately exercising the privilege of sharing their labours; but as a compensation for this, every method must be employed which promotes the multiplication of bees. Thus, for example, if we desire to procure a certain quantity of honey and wax annually, it will be better to seek it in a number of hives, managed with discretion, than to plunder a few of a great proportion of their treasures.

It is indubitable that the multiplication of these industrious animals is much injured by privation of several combs, in a season unfavourable to the collection of wax, because the time consumed in replacing them is taken from that which should be consecrated to the care of the eggs and worms, and by this means the brood suffers. Besides, they must always have a sufficient provision of honey left for winter, for although less is consumed during this season, they do consume some; because they are not torpid, as some authors have conceived[*]. Therefore

---

[*] So far from being torpid in winter, when the thermometer in the open air is several degrees below freezing, it stands at (86) and (88°), in hives sufficiently populous. The bees then cluster together, and move to preserve their heat.

Now that I am on the subject of thermometrical observations, I may cursorily remark, that M. Dubois of Bourg en Bresse, in a memoir otherwise valuable, is of opinion, that the larvæ cannot be hatched below (104). I have repeatedly made the experiment with the most accurate thermometers, and obtained a very different result. When the thermometer rises to (104°), the heat is so much greater than the eggs require, that it is intolerable to the bees. M. Dubois has been deceived, I imagine, by too suddenly introducing his

if they have not enough, they must be supplied with it, which requires great exactness. I admit that in determining to what extent hives may be multiplied in a particular country, it is necessary first to know how many the country can support, which is a problem yet unsolved. It also depends on another, the solution of which is as little known, namely the greatest distance that bees fly in collecting their provender. Different authors maintain, they can fly several leagues from the hive. But by the few observations I have been able to make, this distance seems greatly exaggerated. It appears to me that the radius of the circle they traverse does not exceed half a league. As they return to the hive with the greatest precipitation whenever a cloud passes before the sun, it is probable they do not fly far. Nature which has inspired them with such terror for a storm, and even for rain, undoubtedly restrains them from going so far as to be too much exposed to the injuries of the weather. I have endeavoured to ascertain the fact more positively, by transporting to various distances bees with the thorax painted, that they might again be recognised. But none ever returned that I had carried for twenty-five or thirty minutes from their dwelling, while those at a shorter distance have found their way and returned. I do not state this experiment as decisive. Though bees do not generally fly above half a league, it is very possible they go much farther, when flowers are scarce in their own vicinity. A conclusive experiment must be made in vast arid or sandy plains, separated by a known distance from a fertile region.

Thus, the question yet remains undecided. But without ascertaining the number of hives that any district can maintain, I

---

thermometer into a cluster of bees, and putting them in agitation, the mercury has rose higher than it should naturally do. Had he delayed introducing the thermometer, he would soon have seen it fall to between 95 and 97, which is the usual temperature of hives in summer. In August this year, when the thermometer in the open air stood at 94, it did not rise above 99 in the most populous hives. The bees had little motion, and a great many rested on the board of the hive.

shall remark that certain vegetable productions are much more favourable to bees than others. More hives, for example, may be kept in a country abounding meadows, and where black grain is cultivated, than in a district of vineyards or corn.

◆　　◆　　◆　　◆　　◆

Here I terminate my observations on bees. Though I have had the good fortune to make some interesting discoveries, I am far from considering my labour finished. Several problems concerning the history of these animals still remain unsolved. The experiments I project may perhaps throw some light on them; and I shall be animated with much greater hopes of success, if you, Sir, will continue your counsels and direction. I am, with every sentiment of gratitude and respect,

Francis Huber.

Pregny, 1. October 1791.

APPENDIX.

[The following passages are chiefly engrossed in the substance of the work, but the Translator, as has already been observed, for various reasons, judges it expedient to transfer them to an appendix. In his opinion these very minute details rather interrupt the connexion of the narrative, however interesting they may be considered, and they pertain more to researches purely anatomical.

The Translator has likewise in some instances incorporated several long and important notes with the text; because it appears to him that they actually belong to the substance of the treatise. These are the only variations from the original with respect to arrangement.]

◆ ◆ ◆ ◆ ◆

Swammerdam has given an imperfect description of the ovary of the queen. He observes that he has never been able to find the termination of the oviducts in the abdomen, nor any other parts excepting those which he has described. "Notwithstanding all my exertions, I never could discover the site of the vulva, partly because I had not all my apparatus with me in the country, when investigating this subject, and partly from my apprehension of injuring other parts by pressure, which I had then occasion to examine. However, I have clearly observed a muscular swelling of the oviduct, where approaching the last ring of the belly; that it then contracts and afterwards dilates in becoming membranaceous. As I was desirous of preserving the poison bag, which is situated exactly here, along with, the muscles aiding the motion of the sting, I could follow the oviduct no farther. However, in another female, it appeared that the vulva is in the last ring of the abdomen, and under the sting. The parts expanding only while the queen lays, renders it extremely difficult to penetrate the aperture."

We have attempted to discover what has escaped the indefatigable Swammerdam. But his observation that the research can be made to the greatest advantage, at the time of laying, has paved the way to us. We have remarked that the oviduct did not issue from the body, but that the eggs fall into a kind of cavity, where they are retained several seconds before being laid.

On the sixth of August, we took a very fertile queen, and holding her gently by the wings in a supine position, the whole belly was exposed. She seized the extremity with her second pair of legs, and curved it as much as possible. This seeming an unfavourable position for laying, we forced her to stretch it out. The queen, oppressed with the necessity of laying, could no longer retain her eggs. The lower part of the last ring then separated so far from the upper part as to leave some of the inside discovered. In this cavity the sting lay above in its sheath. As the queen now made new efforts, we saw an egg fall into the cavity from the end of the oviduct. The lips then closed for several seconds; they opened again, and, in a much shorter time, dropped the egg from the cavity.

From our own observations we found that the seminal fluid of drones coagulated on exposure to the air, and from several experiments had so little doubt on the subject, that whenever the female returned with the external marks of fecundation, we thought we recognised it in the whitish substance filling the sexual organs. It did not then occur to us to dissect the females to ascertain the fact more particularly: but this year, whether designing to neglect nothing, or to examine the distension of the female organs, we determined to dissect several. To our infinite surprise, what we had supposed the residue of the prolific fluid, actually proved the genital organs of the male, which separate from his body during copulation, and remain in the female.

We procured a number of queens according to Schirach's method for the purpose of dissection, and set them at liberty that they might seek the males. The first which did so, was seized the instant she returned, and without dissection spontaneously exhibited what we were so impatient to behold. Examining the under part of the belly, we saw the oval end of a white substance which distended the sexual organs. The belly was in constant motion, by alternate extension and contraction. Already had we prepared to sever the rings, and by dissection to ascertain the cause of these motions; when the queen curving her belly very much, and endeavouring to reach its extremity with her hind legs, seized the distending substance with her claws, and evidently made an effort to extract it. She at last succeeded, and it fell before us. We expected a shapeless mass of coagulated fluid; what therefore was our surprise to find it part of the same male that had rendered this queen a mother. At first we could not credit our eyes; but after examining it in every position, both with the naked eye, and a powerful magnifier, we distinctly recognised it to be that part which M. de Reaumur calls the *lenticular* body, or the *lentil*, in the following description.[*]

'Opening a drone there appears a portion formed by the assemblage of several parts, often whiter than milk. This on investigation is found to be principally composed of four oblong pieces. The two largest are attached to a kind of twisted cord, fig. 4. r, called by Swammerdam the root of the penis; and he has denominated seminal vessels, s. s. two long bodies that we are about to consider. Other two bodies oblong like the

[*] *Memoires sur les Abeilles,* p. 450.

preceding, but shorter and not half the diameter, he calls the *vasa deferentia*, d. d. Each communicates with one of the seminal vessels near, g. g. where they unite to the twisted cord, r. From the other extremity proceeds a very delicate vessel, which, after several involutions, terminates in a body, t. a little larger, but difficult to disengage from the surrounding tracheæ. Swammerdam considers these two bodies, t. t. the testicles. Thus there are two parts of considerable size, communicating with other two still thicker and longer. These four bodies are of a cellular texture, and full of a milky fluid, which may be squeezed out. This long twisted cord, r, to which the largest of the seminal vessels is connected, this cord, I say, is doubtless the channel by which the milky fluid issues. After several plications, it terminates in a kind of bladder or fleshy sac, i. i. In different males this part is of various length and flatness. By calling it the *lenticular* body, or the lentil, it receives a name descriptive of the figure it presents in all males whose internal parts have acquired consistency in spirit of wine. The body, l. i. is therefore a lentil, a little thickened, of which one half, or nearly so, of the circumference is edged along the outline by two chestnut coloured scaly plates, e. i. A small white cord, the real edge of the lentil, is visible, and separates them. This lentil is a little oblong, and, for convenience, we shall ascribe two extremities to it, the anterior and posterior. The anterior, l, next the head, is where the canal, r, dividing the seminal vessels is inserted, and the opposite part; i. next the anus, the posterior. The two scaly plates, e. i. e. i, proceed from the vicinity of this last part, whence each enlarges to cover part of the lentil. Under the broadest part of each plate, there is a division formed by two soft points of unequal length; the largest of which is on the circumference of the lentil. Besides these two scaly plates, there are two others, n. n. of the same colour, narrower, and fully one half shorter, each of which is situated very near the preceding, and originates close to the origin of that it accompanies, namely, at the posterior part of the lentil. The rest of the lentil is white and membranaceous. From behind proceeds a tube, k. a canal also white and membranaceous, but it is difficult to judge of its diameter, for the membranes, of which it consists, are evidently in folds. To one side of this pipe is attached a fleshy part, p. somewhat pallet shaped, one side is concave, and the edges plaited; the other side is convex. In certain places the plaits rise and project from the rest of the outline, and form a kind of rays; the pallet appears prettily figured. Though lying with the concave side applied to the lentil, it is not fixed to it. Swammerdam seems to consider this pallet as the characteristic part of the male.

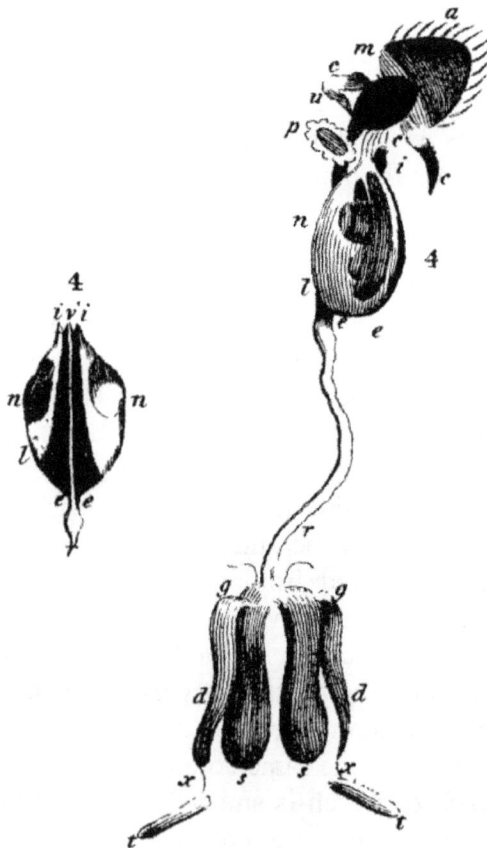

Figure 4

'Though the parts we have described are the most conspicuous in the male, they are neither those which protrude first, nor when protruded are the most remarkable. On viewing from the opposite edge of the lentil, forming the division of the two great scaly plates, a sac or canal, k. proceeding from the posterior part of the lentil, there is distinctly visible the body u, which we call the arc; where there are five transverse hairy bands of a yellow colour, while the rest is white. This arc seems out of the membranaceous canal because it is covered only by a very transparent membrane. One end almost reaches the lenticular body, and the other terminates where the membranaceous canal joins the folded yellow membranes, m. which form a species of sac, that is applied to the sides of the aperture, adapted for the genital organs passing through. These reddish membranes are those that appear first on pressure, and form this elongated portion, at whose end is a kind of hairy mask. Finally, with the sac formed by the reddish membranes, there are connected two appendages, c. c. of reddish yellow, and red at the end, s. These are what appear externally like horns.[*]

The lenticular substance, l. i. provided with each scaly lamina, are the only parts of those described by M. de Reaumur, that we have found engaged in the organs of our queens. The canal, r, by Swammerdam denominated the root of the penis, breaks in copulation; and we have seen its fragments at the place where it unites to the end of the lentil, l. towards the anterior extremity; but we have found no traces of the canal, k, formed of involuted membranes, which in the body of the male proceeds from the posterior end of the lentil, l. i. nor of the plaited pallet, p. adhering to this canal, called by Swammerdam the penis from its resemblance to that of other animals, though he is not of opinion that this point, which is not perforated, can perform the

---

[*] Such long and minute descriptions can be very imperfectly translated; indeed they are unintelligible without microscopical inspections of the parts themselves.—T.

functions of a real penis, and hold the principal part in generation. The canal, k, therefore, and all appertaining to it, must break at i, quite close to the posterior part of the lentil, since we found no remains of the lenticular bodies left by the fecundating males, in the body of our females. The canal, r, which Swammerdam calls the root of the penis, with greater reason than he was himself aware, is not extended in the body of the male as represented by the figure here engraved, but this long twilled canal consists of several involutions, from the seminal vessels whence it proceeds, into the lenticular body where it terminates, and where it conveys the fluid. This canal therefore can extend during copulation, and allow the lenticular substance to protrude out of the body of the males.

It is evident this may be the case during copulation as is seen on opening a drone, for, by endeavouring to displace the lenticular body, the involutions of the cord disappear, and it extends much more than necessary for the lentil to protrude from the body; and if we attempt to separate it farther, the canal breaks at l. close to the lentil, and at the same place where it breaks in copulation.

By dissection two nerves are discovered, towards the origin of the canal, r. inserted into the seminal vessels and distribute in them, and towards the root of the penis many ramifications undoubtedly serving for the motion of these parts. Two small parts, perceptible near the nerves, are two ligaments for retaining the generative organs in their proper place, so that except the root of the penis, they cannot be drawn out without some exertion; it and the lenticular body however can protrude, and actually do so during copulation. A certain degree of pressure forces all these parts from the body of the male, but they spontaneously return, and appear reversed.

Swammerdam, and after him M. de Reaumur have admired this mechanism; they have thought, indeed, that the return should be occasioned by the effect of the air inflating the parts, and they supposed that the male organs proceeded from the body, and returned during copulation, the same as when forced out by pressure. Following their example, we have pressed them from the body of many males; we have a thousand times witnessed this wonderful return, which they detail with the greatest precision; but our males never survived the operation. We have seen, as M. de Reaumur, a few males protrude them spontaneously, even some of the parts inverted, but at that moment they died, and were unable to retract the parts which a pressure, most likely accidental, had forced out. Thus it is improbable that the male organs protrude by turning out of themselves in copulation; and the details which follow prove incontestibly, that it is otherwise. Had not Swammerdam been prejudiced with this opinion, he would have seen that the lenticular body can proceed from the body in erection without reversing itself; he could have proportioned the tortuous canal, which he calls the root of the penis; he would have seen that, at certain times, it can be sufficiently extended to allows the lenticular substance to protrude; he would have discovered the real use of the scaly plates; he would have explained that of the canal k, of the plaited pallet q, and the movements of all these parts, more admirable perhaps than the inversion which he was the first to observe.

Our observations incontestibly prove copulation. The portion of the males found engaged in the body of our queens, hitherto called the lenticular substance, may be denominated a penis both from its position and use. The same surface is presented by it in the queen as in the body of the male, which is proved by the position of the laminæ, e. e. attached to the interior of the penis, when found in the queen. It is evident, if the supposed inversion took place, the laminæ would be found within the posterior part of the penis; and we should see them through its membrane, by their concave side, instead of which the convex surface is presented when in the vulva of females, the same as in the body of the males. But what is the use of these laminæ? From their figure, hardness, relative position with respect to each other, and their situation at the extremity of the penis, we cannot doubt they are real pincers. However, to ascertain the fact, we found it necessary to see their position, and that of the penis itself in the females. For this purpose, we prevented some of the queens from extracting the parts left by the impregnating males, and by dissection we discovered that the laminæ were pincers as we had conjectured.

The penis was situated under the sting of the queens, and pressed against the upper region of the belly. It was supported by the posterior end, against the extremity of the vagina, or excretory canal. There we were sensible of the motion and use of the scaly pieces. Their extremities were separated a little more than in the male, and pressed between them some of the female parts below the excretory canal. The extreme minuteness of these parts prevented us from distinguishing them clearly, but the effort necessary to separate and remove the penis from the female, satisfied us of the use of these laminæ.

Inspecting a male from above, the convex side of the plates, e. e. is presented, and the summit of the angle formed by their origin. When in the body of the female, they are in the inverse position; what was above in the male is now below, and the extremity of the pincers directed upwards. This makes us suspect that in copulation the male mounts on the back of the female, but we are far from asserting it positively. It may be asked whether that part we call the penis, is the sole part introduced into the female during copulation? We have carefully investigated this, and can affirm, that it is the only one of all those described by M. de Reaumur, which has been found in our females. But we have discovered a new part that escaped both him and Swammerdam, which appears from the following experiment.

Separating the lenticular substance from the excretory canal, where it was attached, we drew along with it a white body, adhering by one extremity, and having the other engaged in the vagina. Towards the end of the lentil, where the substance adhered, it appeared cylindrical, then it swelled, and again contracted, to dilate anew in a greater degree than at first; afterwards it contracted and terminated in a point. A powerful magnifier was required to see all this. When pulled from the lenticular body, the part was commonly broke, and also when extracted by the queens from themselves. The figure and situation seemed to authorise our considering it the penis itself, and the lenticular body only an appendage. But the last queen we examined exhibited a peculiarity that induced us to doubt the fact, and led us to suspect that this body is nothing else than the seminal fluid itself, moulded and coagulated in the vagina, and which from its viscosity adheres to the lenticular substance, and accompanies it when separated from the vagina. In this queen was found a little extravasated white matter, near the opening of the vagina. This, though at first liquid, soon coagulated in the air as the seminal fluid of drones does. In separating the lenticular body from the vagina, we drew along with it a thread which broke near the lentil; and seemed of too little consistence for the penis of a male. The lenticular bodies, found in our queens, appeared larger than in the males we dissected, and we have remarked with M. de Reaumur, that these parts are not of equal size in every male.

Experiment 1.—On the tenth of July, we set successively at liberty three virgin queens four or five days old. Two flew away several times; their absence was short and fruitless. The third profited better by her liberty; she departed thrice; the first and second time her absence was short; but the third lasted thirty-five minutes. She returned in a very different state; and in such as allowed no doubt of her employment, for she exhibited the part of a male that had rendered her a mother. We seized her wings with one hand, and in the other received the lenticular body, of which she had disengaged herself with her claws. The posterior part was armed with two pincers, e. e. shelly and elastic, which could be drawn asunder, and then resumed their original position. Towards the anterior part of the lentil appeared the fragment of the root of the penis; this canal had broke half a line from the lenticular body. We allowed the queen to enter her habitation, and adapted the entrance so that she could not leave it unknown to us.

On the seventeenth we found no eggs in the hive; the queen was as slender as the first day; therefore the male, with which she had copulated, had not impregnated her eggs. She was again set at liberty; after twice departing, she returned with evidence of a second copulation. We then confined her, and the eggs she afterwards laid proved that the second copulation had been more successful than the first and that there are some males more fit for impregnating queens than others. However, it is very rare that the first copulation is inefficient; we have only seen two that required it twice; all the rest were impregnated by the first.

Experiment 2.—On the eighteenth we put at liberty a virgin queen twenty-seven days old, she departed twice. Her second absence was twenty-eight minutes, and she returned with the proofs of copulation. We prevented her from entering, and put her under a glass to see how she would disengage the male organs. This she was unable to accomplish, having only the table and sides of the glass for support; therefore we introduced a bit of comb; thus providing the same conveniences as are in a hive. Fixing herself on *it* by the first four legs, she stretched out the two last, and extending them along her belly seemed to press it between them. At length introducing her claws between the two parts of the last ring, she seized the lenticular body, and dropped it on the table. The posterior part was provided with shelly pincers, under which and in the same direction was a grey cylindrical body. The end farthest from the lentil was sensibly thicker than that adhering to it, and terminated in a point. This

148

point was double, and open like the bill of a bird, which induces us to think the body was broken, a conjecture supported by the following experiment.

Experiment 3.—On the nineteenth we set at liberty a queen four days old; she departed twice; her first absence was short; the second lasted thirty minutes, and then she returned with the marks of fecundation. As we wished to obtain the male organs entire, it was necessary to prevent the queen from breaking them by extracting them with her feet; we therefore suddenly killed her, and cut off the last rings in order to lay the vulva open. But though deprived of animation, so much life remained in these parts that the lenticular body was thrown out spontaneously. Under the pincers appeared the remnant of a cylindrical body which had broken near the origin and remained in the female. This body was very small at the origin; it afterwards sensibly enlarged; next contracting by degrees, it terminated in a sharp point. We found the point engaged up to the gland in the excretory canal, and the rest in the vulva.

Experiment 4.—We set two virgin queens at liberty on the twentieth. The first had been abroad on the preceding days, but the scarcity of males prevented her from being previously fecundated. She returned with the organs of a male. We tried to prevent her from extracting them, but she did this so expeditiously with her feet, that we could not accomplish it. She was then allowed to enter the hive.

The second queen departed twice. Her first absence was short as usual; the second lasted about half an hour, and she returned impregnated. Having killed her as suddenly as possible, we laid open the vulva. The lenticular body was deposited as in every queen hitherto dissected; the pincers were situated under the excretory canal. Some parts not easily distinguishable were pressed between the laminæ, and their office seemed to consist in forcing the extremity of the lentil to approach the orifice of the vagina, and apply so forcibly to it that some exertion was necessary to separate them. We previously examined them, with a very powerful magnifier. Then a peculiarity which had escaped us was perceptible. In drawing out the lenticular body, there proceeded from the vagina a minute part, v. adhering to the posterior end of the lentil, and situated below the plates. It spontaneously retracted into the lentil, like the horns of a snail. It appeared white, very short, and cylindrical. Under the pincers was a little half coagulated seminal fluid at the bottom of the vulva. Though much could be expressed, there was none pure; it was almost liquid, but soon coagulated, and formed a whitish inorganic mass. This observation carefully made removed all our doubts, and demonstrated that what we had taken for the penis of males was nothing but the seminal fluid, which had coagulated and assumed the interior figure of the vagina. The only hard part introduced by the male, was the short cylindrical point which retracted into the lentil, when we separated it. Its situation and office prove that it is there we must look for the issue of the seminal fluid, if we can hope to find an opening, when not engaged in copulation.

We found this new part in the first drone we dissected. By pressing the seminal vessels, the white liquid then escaped downwards to the root of the penis r. and into the lenticular body, l. i. which became sensibly swoln. We prevented the fluid from returning, and by new pressure of the lentil forced it to advance. However, none escaped, but we saw at the posterior end of the lenticular body, and under the scaly pincers, a small white cylindrical substance, the same in appearance as that we had found engaged in the vagina of the queen. This part retracted on pressure, and then returned.

I request you, Sir, while perusing this letter, to inspect the figure of the male sexual organs published by M. de Reaumur, and which are copied here. The descriptions are most accurate, and present a just idea of the situation of these parts when in the

150

male's body. We readily conceive how they appear when left in the female by copulation. This detail will sufficiently indicate the situation and figure of the new part I have discovered.

I suspect that the males perish after losing their sexual organs. But why does nature exact so great a sacrifice? This is a mystery which I cannot pretend to unveil. I am unacquainted with any analogous fact in natural history, but as there are two species of insects whose copulation can take place only in the air, namely, ephemeræ and ants, it would be extremely interesting to discover whether their males also lose their sexual parts, in the same circumstances, and whether, as with drones, enjoyment in their flight is the prelude of death.

FINIS.

ANALYTICAL INDEX.

Also from Benediction Books ...
Wandering Between Two Worlds: Essays on Faith and Art
Anita Mathias
Benediction Books, 2007
152 pages
ISBN: 0955373700

Available from www.amazon.com, www.amazon.co.uk

In these wide-ranging lyrical essays, Anita Mathias writes, in lush, lovely prose, of her naughty Catholic childhood in Jamshedpur, India; her large, eccentric family in Mangalore, a seacoast town converted by the Portuguese in the sixteenth century; her rebellion and atheism as a teenager in her Himalayan boarding school, run by German missionary nuns, St. Mary's Convent, Nainital; and her abrupt religious conversion after which she entered Mother Teresa's convent in Calcutta as a novice. Later rich, elegant essays explore the dualities of her life as a writer, mother, and Christian in the United States-- Domesticity and Art, Writing and Prayer, and the experience of being "an alien and stranger" as an immigrant in America, sensing the need for roots.

About the Author

Anita Mathias is the author of *Wandering Between Two Worlds: Essays on Faith and Art.* She has a B.A. and M.A. in English from Somerville College, Oxford University, and an M.A. in Creative Writing from the Ohio State University, USA. Anita won a National Endowment of the Arts fellowship in Creative Nonfiction in 1997. She lives in Oxford, England with her husband, Roy, and her daughters, Zoe and Irene.

Visit Anita's website
   http://www.anitamathias.com,
and Anita's blog
   http://dreamingbeneaththespires.blogspot.com (Dreaming Beneath the Spires).
.

The Church That Had Too Much
Anita Mathias
Benediction Books, 2010
52 pages
ISBN: 9781849026567

Available from www.amazon.com, www.amazon.co.uk

The Church That Had Too Much was very well-intentioned. She
wanted to love God, she wanted to love people, but she was both
hampered by her muchness and the abundance of her possessions, and beset by ambition, power struggles and snobbery.
Read about the surprising way The Church That Had Too Much
began to resolve her problems in this deceptively simple and en-
chanting fable.

About the Author

Anita Mathias is the author of *Wandering Between Two Worlds:
Essays on Faith and Art.* She has a B.A. and M.A. in English
from Somerville College, Oxford University, and an M.A. in
Creative Writing from the Ohio State University, USA. Anita
won a National Endowment of the Arts fellowship in Creative
Nonfiction in 1997. She lives in Oxford, England with her husband, Roy, and her daughters, Zoe and Irene.

Visit Anita's website
   http://www.anitamathias.com,
and Anita's blog
   http://dreamingbeneaththespires.blogspot.com (Dreaming Beneath the
Spires).

www.ingramcontent.com/pod-product-compliance
Lightning Source LLC
Chambersburg PA
CBHW022110280326
41933CB00007B/325